SpringerBriefs in Water Science and Technology

More information about this series at http://www.springer.com/series/11214

Andrea Guerrini · Giulia Romano

Water Management in Italy

Governance, Performance, and Sustainability

Andrea Guerrini
Management
University of Verona
Verona
Italy

Giulia Romano
Economics and Management
University of Pisa
Pisa
Italy

ISSN 2194-7244
ISBN 978-3-319-07817-5
DOI 10.1007/978-3-319-07818-2

ISSN 2194-7252 (electronic)
ISBN 978-3-319-07818-2 (eBook)

Library of Congress Control Number: 2014943937

Springer Cham Heidelberg New York Dordrecht London

© The Author(s) 2014

This work is subject to copyright. All rights are reserved by the Publisher, whether the whole or part of the material is concerned, specifically the rights of translation, reprinting, reuse of illustrations, recitation, broadcasting, reproduction on microfilms or in any other physical way, and transmission or information storage and retrieval, electronic adaptation, computer software, or by similar or dissimilar methodology now known or hereafter developed. Exempted from this legal reservation are brief excerpts in connection with reviews or scholarly analysis or material supplied specifically for the purpose of being entered and executed on a computer system, for exclusive use by the purchaser of the work. Duplication of this publication or parts thereof is permitted only under the provisions of the Copyright Law of the Publisher's location, in its current version, and permission for use must always be obtained from Springer. Permissions for use may be obtained through RightsLink at the Copyright Clearance Center. Violations are liable to prosecution under the respective Copyright Law.

The use of general descriptive names, registered names, trademarks, service marks, etc. in this publication does not imply, even in the absence of a specific statement, that such names are exempt from the relevant protective laws and regulations and therefore free for general use.

While the advice and information in this book are believed to be true and accurate at the date of publication, neither the authors nor the editors nor the publisher can accept any legal responsibility for any errors or omissions that may be made. The publisher makes no warranty, express or implied, with respect to the material contained herein.

Printed on acid-free paper

Springer is part of Springer Science+Business Media (www.springer.com)

Contents

1 Introduction 1

2 **The Italian Water Industry** 5
   2.1 A Brief Overview of Italian Water Reforms: A Twenty-Year Excursus 5
   2.2 The Current Regulatory Framework 10
   2.3 An Overview of the Italian Water Industry 12
   References 15

3 **The Determinants of Water Utilities Performance** 17
   3.1 The Effects of Ownership and Political Connections on the Performance of Water Utility Companies: An Overview 17
   3.2 Boosting Efficiency Through Economies of Scale, Scope, and Population Density: Evidence from Prior Studies 20
   3.3 Data Collection and Research Method 22
      3.3.1 Data Collection and Description 22
      3.3.2 DEA Analysis 25
      3.3.3 Statistical Analysis 27
   3.4 Results and Discussion 28
   3.5 Conclusions 31
   References 32

4 **Investments Policies and Funding Choices** 37
   4.1 Investments Realization and Infrastructures Needs 37
   4.2 Factors Limiting the Investment Realizations in the Italian Water Sector: The Experience of Acque Veronesi s.c.a r.l 40
   4.3 Investment Policies and Funding Choices in the Water Sector: The Need of an Empirical Survey 44
   4.4 Data Collection and Method of Analysis 46
   4.5 Results and Discussion 50
   References 53

5 Water Demand Management and Sustainability ... 55
5.1 Sustainable Use and Management of Water Resources: A Brief Overview ... 55
5.2 Policies for Sustainable Water Use: A Review of the Literature ... 57
5.2.1 Tariff Policy ... 57
5.2.2 Rationing and Restrictions ... 63
5.2.3 Technology Devices ... 66
5.2.4 Information Campaigns ... 70
5.3 Promoting Conservation Practices of Water Use Through Web Sites: An Empirical Analysis on Italian Water Utilities ... 71
5.3.1 Data Collection and Method Adopted ... 71
5.3.2 Results and Discussion ... 72
5.4 Wastewater Technologies to Reduce Environmental Impacts ... 75
References ... 79

6 Conclusions ... 85
References ... 88

Contributors

Bettina Campedelli Full Professor at the Department of Business Administration—University of Verona

Francesco Fatone Assistant Professor at the Department of Biotechnology—University of Verona

Luciano Franchini Engineer, is main director (general manager) of Consiglio di Bacino Veronese, local authority for regulation of integrated water services, since 2002

Andrea Guastamacchia Chief Financial Officer of Acque Veronesi s.c.a r.l., a full-publicly owned water utility

Andrea Guerrini Assistant Professor at the Department of Business Administration—University of Verona

Martina Martini Ph.D. student at the Department of Business Administration—University of Verona

Giulia Romano Assistant Professor at the Department of Economics and Management—University of Pisa

Giorgia Ronco At present serves as officer at the National Authority for Energy, Gas and Water Systems (AEEGSI); her contribution was developed before her engagement with AEEGSI and reflects the sole personal opinion of the author

Chapter 1
Introduction

Water is essential for life and for the economy and is one of the main environmental topics of European Union (EU) policy. Even if most Europeans have historically been shielded from the social, economic, and environmental effects of severe water shortages, the gap between the demand for and availability of water resources is reaching critical levels in many parts of Europe. Climate change is likely to exacerbate current pressures on European water resources. Moreover, much of Europe will increasingly face reduced water availability during the summer months, and the frequency and intensity of drought is projected to increase, particularly in the southern and Mediterranean countries. Thus, the EU is showing increasing concern regarding drought events and water scarcity, and policymakers and utility managers must face the challenge of balancing the increasing human demand for water with the protection of ecosystem sustainability. The Water Framework Directive (2000/60/EC), the most relevant European Water Framework, is based on the idea that water management needs to take account of economic, ecological, and social issues and that its prime objective is the sustainable use and management of water resources.

In Italy as in many other countries, an intense debate over the water industry is ongoing. Policymakers are looking for the most effective strategies for efficient water management, focusing on governance and organizational choices. Italy is facing many problems in terms of the technical efficiency, economic profitability, and financial sustainability of its water utilities as well as water scarcity and inefficient water use, since leakages accounted for around 36 % of the water fed into Italy's water grid, with an average maximum of 43 % in the south.

According to Eurostat data, Italy's total freshwater abstraction by public water supply is the highest in Europe. The Italian unit price of household water supply and sanitation services is among the lowest among Organization for Economic Cooperation and Development (OECD) countries, though it has increased, rising an average of 5 % from 2007 to 2008 and 6 % from 2004 to 2008. Moreover, due to low tariffs, water consumption in Italy is still higher than in other European countries. Italy's household water consumption in 2002 was 206 l, with a decrease over the subsequent 10 years of around 15 % (Istat 2013). However, data show that the average water usage per person in Italy is the highest among European countries.

© The Author(s) 2014
A. Guerrini and G. Romano, *Water Management in Italy*, SpringerBriefs in Water Science and Technology, DOI 10.1007/978-3-319-07818-2_1

This book is an attempt to discuss the most relevant issues concerning water management in Italy. Using the most recent available data and starting from the extant international literature, it focuses on the features of the Italian water industry, the water utility firms' efficiency, the investment policies and funding choices of the water companies, and the sustainable practices put in place by the utilities to reduce water consumption and spread virtuous behaviors.

Chapter 2 analyzes the evolution of Italy's legal framework, starting from the first relevant water reform in 1994, then highlighting the current framework. Starting with the full list of Italian institutions (1,235 firms and public bodies) operating in the water industry, it then analyzes the main features of the industry by collecting data from the National Authority for Energy, Gas and Water Services (AEEG) database, the Bureau Van Dijk AIDA database, financial statements, and corporate websites on institution type, geographical localization, and water services provided (i.e., collection, potabilization, adduction/transportation, distribution of water for civil use, sewerage, and wastewater treatment).

Focusing on 304 water firms, the study examines diversification strategies (either mono or multiutilities), firm size (considering the number of employees), ownership type (i.e., public, private or mixed-ownership), and number of shareholders.

Chapter 3 focuses on the determinants of Italian water utility performance. Starting with a literature review on the effects of ownership and political connections on firm results and the existence of economies of scale, scope, and population density, the chapter empirically studies the factors affecting the performance of 98 mono-utility water companies involved in integrated water services (the simultaneous provision of all of the main water services—collection, adduction/transportation, distribution of water for civil use, sewerage, and wastewater treatment) covering a period of 5 years (from 2008 to 2012).

Performance was assessed through the Data Envelopment Analysis Method in order to describe firm efficiency; then, statistical analyses were conducted to determine whether firm size, customer density (measured as the ratio of population served to kilometer of main length), geographical localization (north, center, or south) and ownership (public or mixed-private) are relevant factors affecting firm efficiency.

Chapter 4 examines the investment policies and funding choices of Italian water utilities. Starting with a description of the investments needed to improve water services, implement new technologies, and reduce water leakages and waste, the chapter reveals the differences between expected investments in the last few years and the investments effectively realized.

The chapter analyzes the factors limiting the investment realizations in the Italian water sector by examining the case of Acque Veronesi s.c.a r.l., a medium-sized utility operating in Veneto, in the north of Italy.

Finally, after a comprehensive literature review on this issue, an empirical study using information on the abovementioned 98 mono-utility water companies covering 2008–2012 is conducted. Considering the relevant financial indicators of investment and funding choices, the study enquires if firm size, customer density, geographical localization, and ownership are significant factors affecting firm decisions.

Chapter 5 provides an overview of the increasingly important issue of the sustainable use and management of water resources. It reports a literature review of the most relevant studies on the implementation and effectiveness of the instruments used by water utilities to implement water conservation policies (such as water pricing, incentives for the implementation of high-efficient appliances, rationing policies, and information campaigns to improve awareness of activities useful in reducing water consumption). Moreover, it provides an empirical analysis of the willingness of Italian water utilities to provide through their corporate websites information about reducing household water consumption and the water quality they provide to customers. Finally, it describes the wastewater technologies used to reduce environmental impacts.

The book concludes with an analysis of the most pertinent strengths, weaknesses, opportunities, and threats facing the water industry in Italy, with the aim of providing policymakers, decision leaders, utilities managers, and interested citizens a comprehensive framework for informing later steps in water management and achieving the objective of offering the proper attention, the necessary economic resources, and the required commitment to solve the "water issue."

The book aims to contribute to the current EU environmental policy mainstream focusing on the need to reconcile the triple objectives of wealth creation, social cohesion, and environmental protection, being aware that "scientifically sound tools to support decision-making by measuring and assessing policies' impact are needed for the successful implementation of genuinely sustainable policies."[1]

Last but not least, we deeply thank all the researchers and practitioners who have contributed to the development of this book, created as an attempt to effect a close collaboration between university researchers, utility managers, and policymakers in compliance with the EU Horizon 2020 framework stressing the importance of cooperation between the public and private sectors and between universities and business.

Reference

Istat, Istituto Nazionale di Statistica (2013) Italia in cifre. http://www.istat.it/it/files/2011/06/Italia_in_cifre_20132.pdf

---

[1] http://ec.europa.eu/research/environment/index_en.cfm?pg=tools.

Chapter 2
The Italian Water Industry

2.1 A Brief Overview of Italian Water Reforms: A Twenty-Year Excursus

The Italian Integrated Water Supply system presents a very complex landscape. Italy's water main and wastewater treatment plant network is very heterogeneous. Best practices exist, where entire areas are fully served by drinking water flowing directly to their homes all day, but there are other areas where the water flows from the tap only a few days a week. Municipalities served with high-quality water by innovative technologies coexist with poor areas characterized by outdated mains providing low-quality water.

The same applies to the sewerage systems and, above all, the treatment plants. There are many efficient and innovative wastewater treatment plants and many plants built years ago and now abandoned or poorly maintained. The European Community (EU) has begun several infringement proceedings against Italy, as it is not meeting the deadlines for the transposition of EU directive 271/91 for wastewater: the terms of adoption have long expired. In 2012, the European Commission took Italy to the EU Court of Justice for its failure to ensure that wastewater from agglomerations with more than 10,000 inhabitants discharging into sensitive areas is properly treated. In 2011, the Commission informed Italy that over 143 towns were still not connected to a suitable sewage system and/or lacked secondary treatment facilities or had insufficient capacity. While considerable progress has been made, 14 years after the deadline expired (in 1998, as the EU legislation required), at least 50 agglomerations still have shortcomings. The Commission claimed that the lack of adequate collection and treatment systems poses risks to human health and to inland waters and the marine environment.[1]

Sections 2.1, 2.2 were written by Bettina Campedelli, Luciano Franchini and Giulia Romano, while 2.3 was written by Giulia Romano.

[1] http://europa.eu/rapid/press-release_IP-12-658_en.htm.

The Italian water industry needs to provide the investments required to address this critical situation. If we consider the infrastructure needs for the entire water supply sector, the total volume of investments needed reaches € 64 billion (D'Angelis and Irace 2011). However, the scarcity of funds available to national and local governments and the effects of the EU Stability and Growth Pact limit municipalities' investment capacity for water infrastructure and service improvements.

Attracting private investment could offer a solution, though investors are not currently interested in the Italian water sector because of its unstable legal framework (which has rapidly changed in the last 7 years) and the need to dialog with the local governments and politicians who manage a large part of the industry.

Although regulation of the Italian water industry began in 1865 (Marques 2010), the most comprehensive reform of water sector regulation began in the 1990s. In 1994, the Italian Parliament enacted the first law for the reorganization of the integrated water service (SII) in response to the emergency situation affecting a large part of the country. The SII covers the public collection, transportation, and distribution of water for civil use as well as sewerage and wastewater treatment for both mixed-use residential and industrial clients.[2]

Law 36/1994 (called the "Galli law," for Giancarlo Galli, the Italian parliamentarian who was its principal author) tried to reorganize water services management, promoting the elimination of all direct municipal management and all the micro-enterprises that remain part of the Italian water system.

The Galli law was approved in 1994 and then applied along with subsequent regulations, such as ministerial rule 01/08/1996 on tariffs (the so-called "Normalized Method") and law 152/2006 (the so-called "Environmental Code"). The main principles of the Galli law are the following:

- Surface water and groundwater, although not extracted from the subsoil, are public and must be maintained and used in accordance with the criteria of equity;
- Any use of water must safeguard the expectations and rights of future generations, so that they will benefit from a well-preserved natural heritage;
- Water use will follow the principles of water savings and renewal and must not affect water resources, the liveability of the environment, agriculture, fauna and aquatic flora, geomorphological processes, and hydrogeological equilibrium;
- Water use for human consumption has priority over other types of use, which are allowed when the resource is sufficient and preserving the quality of water for human consumption is possible.

The law aimed to overcome the permanent emergency affecting the integrated water services and promote the conditions for effective regulation of the industry. It provides, in the medium term, full water services coverage for the entire population and environmental protection through the construction of new sewers and wastewater treatment plants.

---

[2] National Authority for Energy, Gas and Water Services (AEEG).

Afterward, industrialization started to incentivize mergers and aggregations among utilities: large and diversified firms are best able to collect the necessary funds to cover all operating costs and finance infrastructure investments. In other words, the new law induced firms to try to produce economies of scale and scope by achieving cost efficiencies.

The law delegated to the regions the duty of identifying "optimal areas" (*Ambito Territoriale Ottimale*, or "ATO") to be managed under the supervision of a local public authority for water services (*Autorità d'Ambito Territoriale Ottimale*, of "AATO"); however, though some regions quickly complied with the law (such as Tuscany and Lazio, which defined their ATOs in 1995 and 1996), other regions waited a long time to define theirs.

Law 36/1994 decrees that the management of the SII can occur under a private company, mixed-ownership company, or public company. In the case of a direct award to companies totally publicly owned or with a majority of public shares, an AATO, may entrust water services without recourse to competitive tendering. Otherwise, the AATO must conduct competitive tendering.

In order to maintain efficiency, effectiveness, and cost-effectiveness, local governments may provide for the management of the SII through a plurality of firms (e.g., one firm may provide the distribution and another the wastewater and sewerage).

In entrusting water management to an industrial company, a local authority negotiates with the concessionaires the required standards of service quality and investment needs. The execution of the plan and the service delivery are the utility's responsibilities, while the municipality must periodically monitor activities through the AATO.

The Galli Law provided for the establishment of a tariff system based on the principle of a single tariff for each ATO, including the drinking water supply, sewerage, and waste water, to ensure full coverage of the operating costs and investment. The tariff is determined taking into account a variety of factors, including the quality of the water resource and the service provided, the investment and necessary maintenance, the extent of the operating costs, and the adequacy of the return on investment. These factors must all be weighed in relation to the financial plan for the investments: the tariff is determined on the basis of the "reference tariff," used to adjust the tariff over time. To do this, the AATO takes into account the objectives of improved productivity and service quality and the current rate of inflation.

On August 1, 1996, the Minister of Public Works established the so-called "Normalized Method" to define the cost components and determine the reference tariff.

The Galli law confers significant autonomy onto each local authority, empowering AATOs to reorganize and oversee the water system. However, the law generated a high level of heterogeneity across the country, allowing many different ways of arranging water services.

In sum, law 36/1994 is a general framework that needed further regulations to be effectively applied; it provides no standards for delegating water services management, which is left to the regions and local authorities.

A further limitation of the 1994 reform was its lack of an independent regulatory authority for water. In the beginning, supervision was carried out by a committee of the Ministry of Public Work, which was transformed into the Commission of the Ministry for Environment (*Commissione Nazionale per la Vigilanza delle Risorse Idriche*, or the Co.N.Vi.Ri). Both entities were closely linked to the government and lacked the autonomy and independence they needed.

Moreover, the 2000 Water Framework Directive established a framework for EC action on water policy. The Directive introduces two key economic principles: it calls on water users (i.e., households, industries, and farmers) to pay for the full costs of the water services they receive and on Member States to use economic analyses in the management of their water resources and assess both their cost-effectiveness and the costs of alternatives when making key decisions.[3]

Twenty years ago, Italy had an opportunity to reform its national water sector, but this goal has been only partially achieved. After the promulgation of the Galli law, many areas of the country remain without effectively organized water services. Thus, 20 years after the reform went into effect, its purpose has not been completely achieved, though progress has been made: many firms now integrate their water, wastewater, and sewerage services (Co.N.Vi.Ri 2009), and some are now multiutility, providing services for the gas, electrical energy, and waste industries.

Further legislative interventions occurred over the last 20 years, but they were not completely consistent with each other and did not substantially improve the sector's organization.

Twelve years after the Galli Reform, Law 152/2006 provided new standards for the organization and control of water services. It regulates the water sector in an organic way, incorporating Law 36/94 and dictating more precisely the tasks and activities relevant to the various institutional actors involved in the water industry. Under the new law, the AATOs are now defined uniformly across the country instead of according to regional regulations.

Law 152/2006 defines the powers and responsibilities within the water sector as follows:

1. A National Regulatory Authority should define the national framework under which all firms must operate, choosing the tariff method and the service contract type; then, it should periodically monitor the implementation of the rules in every area.
2. A Local Regulator Authority (AATO) is responsible for controlling the entities that locally manage the services.
3. An entrusted water utility company is the owner of service delivery and the implementation of the necessary infrastructure.

---

[3] http://ec.europa.eu/environment/water/participation/pdf/waternotes/water_note5_economics.pdf.

2.1 A Brief Overview of Italian Water Reforms: A Twenty-Year Excursus

The relationship among these three actors is characterized by an intense reporting flow. Every AATO draws up a plan of the structural and organizational changes required to achieve the water and service quality targets established through national law and negotiated in detail with the utilities. This document is then matched with a business plan that includes an income statement, an asset and liability statement, a cash flow statement, and the financial ratios for each year covered by the license. Both documents are periodically revised and sent to the National Authority for Energy and Gas (AEEG) for approval. A third document, called the "contract of service," negotiated between the AATO and the utilities, defines the standard of services and identifies the key performance indicators the local regulatory authority must monitor. Water services might be entrusted to:

- a private company chosen through a public competitive tender;
- mixed-ownership company, the private partner of which is chosen through a public competitive tender;
- public company, with an in-house provision of services.

The decree of January 16, 2008, n. 4, changed Law 152/2006, particularly to admit more entrusted water utility companies to the same ATO.

A map of Italian ATOs was designed by regional local authorities to chart the hydrological basins and the administrative boundaries. The map's divisions were intended to create large areas that could be financially self-sufficient through tariff collections.

Figure 2.1 shows the 2009 distribution of Italian ATOs. The most common service cluster is between 250 and 400,000 inhabitants; however, quite a few ATOs operate in the lower and upper clusters (20 and 24 ATOs, respectively).

In 2008, nearly 15 years after the Galli reform, its planned changes had still not been fully achieved, despite certain improvements. The last report of the Co.N.Vi.Ri showed that, in 2008, only 75 % of AATOs had finished reorganizing

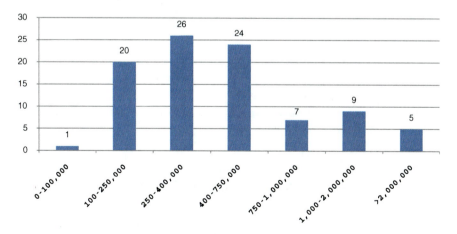

Fig. 2.1 Number of ATOs and size of population served. *Source* (Co.N.Vi.Ri 2009)

and franchising water and wastewater services to independent firms, serving 57 % of Italian municipalities and 66 % of the Italian population; in the remaining municipalities, most water services were still being provided by the municipalities (Co.N.Vi.Ri 2009 and 2011).

In response to this situation, the Italian government mandated the privatization of public services, including water and wastewater services (by modifying Law 133/2008, article 23 bis in November 2009). The intent of this reform is to improve SII performance through the introduction of private investors whom the Italian government considers to be more oriented toward efficiency and effectiveness than public investors are. Under this new reform, water and wastewater services had to be franchised to private or public–private utilities in which the private partner held at least 40 % of the shares; no water management franchises could be awarded to totally publicly owned utilities after December 2011 (Testa 2010).

This change prompted extensive political debate in Italy among a large part of the population: those in favor of water industry privatization believed that the private provision of water services would improve quality and efficiency and thus reduce tariffs, while supporters of public water systems were convinced that water services should not be privatized, being a natural monopoly, and that private players would not improve investments or water quality but only increase their profits. Moreover, they criticized the existing tariff system that allowed a 7 % assured return on invested capital even for inefficient firms (Guerrini and Romano 2013).

Two 2001 referenda on these issues attracted broad public participation. The outcome was that AATOs were no longer obliged to franchise water and wastewater services only to mixed or privately owned utilities; they could grant concessions to public companies financed by municipalities, as they could before the 2009 reform. In addition, the tariff-setting method changed: water tariffs no longer had to guarantee a return on invested capital.

The 2010 Law n. 42 mandated the deletion of the AATOs not later than January 1, 2011 (later extended to December 31, 2012), conferring the AATO's functions onto the regions through a new law. The number of AATOs dropped to 71, since four Italian regions (Emilia Romagna, Tuscany, Abruzzo, and Calabria) opted for unique regional AATOs. In Tuscany, for example, instead of six different AATOs, the A.I.T. (*Autorità Idrica Toscana*) has operated alone since the beginning of 2012.

In 2011, Law 214/2011 gave the AEEG the power to supervise the water sector, in addition to the gas and energy sectors it already regulated. The AEEG is governed by a committee of five members who sit for 7 years; each member is named by the Italian government and then approved by parliamentary committees, and they represent all the major political parties.

2.2 The Current Regulatory Framework

As reported in the previous paragraph, the current regulatory framework is the result of the many attempts to liberalize and modernize the SII made by various governments over the last two decades (Guerrini and Romano 2013;

2.2 The Current Regulatory Framework

Carrozza 2011; Danesi et al. 2007). It is also the result of the European framework drawn through the Water and Waste Water Directives (Directive 2000/60/EC and Directive 91/271/EEC) and the overwhelming majority in June 2011 public referendum that delayed compulsory water services privatization and the guaranteed return on investment for water utilities.

Law 152/2006, the Water Framework Directive, and decree n. 201/2011 comprise the current national framework for water services. The latter decree conferred the regulation and control of water services onto the AEEG, with the Ministry of the Environment responsible for other functions (e.g., defining the general objectives of water quality, developing ways to encourage water conservation, water use efficiency, and wastewater reuse). The AEEG regulates water services according to the following aims:

- guaranteeing the dissemination, accessibility and quality of services to users uniformly throughout the country;
- establishing a tariff system that is fair, reliable, transparent, and non-discriminatory;
- protecting the rights and interests of users;
- managing water services in terms of efficiency and economic and financial stability;
- implementing the European Community's "full cost recovery" (including environmental and resource-related costs) and "the polluter pays" principles.

To achieve these aims, the AEEG defined a tariff method for determining the rate of water service, paying particular attention to reimbursing operating costs, service costs, and the related environmental costs of the resources. The Authority began its activities in 2012 by issuing a transitional tariff model (MTT) and then developed a new model (the *Metodo Tariffario Idrico*, or MTI) that is more consistent with EU standards and respectful of the outcome of 2011 referendum. The MTT replaces the model that had been in force since 1996 and was applied in 2012 and 2013 before being replaced with the MTI in 2014. It is worth briefly explaining the MTI, since it affects businesses significantly. The new pricing formula is as follows:

$$VRG^a = Capex^a + FoNI^a + Opex^a + ERC^a + Rc^a_{TOT}$$

where:

- Capex: represents the cost of fixed assets, including interest expenses, tax expenses, depreciation, and amortization;
- FoNI: includes cost items paid to finance new investments;
- Opex: includes operating costs;
- ERC: covers the environmental and resource costs not included in the other tariff components;
- Rc: represents adjustments for the prior years' tariff.

The MTI provides a new paradigm for tariff estimation: the previous "normalized method" was based on *ex-ante regulation*, which determines a tariff on the basis of planned investments; the MTI applies CAPEX tariff coverage through an *ex-post regulation* that includes only those costs related to actual investments. The new model thus transfers the risk of delayed returns on investment from the citizens to the water utilities.

This provision represents a significant reform that could improve the quality of services. The former method did not incentivize firms to realize their investments, as they were reimbursed for the cost of their planned investments even when not realized. Under the *ex-ante regulation*, several utilities experienced high tariffs and low investments (Guerrini et al. 2011). In such cases, the AATO sanctions the firms, but the authority does not often exert effective control. The MTT and MTI will be further described in Chap. 4.

2.3 An Overview of the Italian Water Industry

A recent survey (AEEG 2013) on a sample of 284 water utilities shows that Italy has highly heterogeneous service area sizes (see Table 2.1). The average number of municipalities served by a single firm is 12, highlighting the severe fragmentation of the Italian water industry. This is shown in Fig. 2.2, which indicates that 117 out of 284 selected firms operate in an area with fewer than 5,000 inhabitants.

Many firms are still operating on limited hydrological basins. Moreover, some municipalities have not yet delegated the management of their water services, which furthers the aggregation and corporatization of the Italian water sector.

The AEEG database indicates that 1,235 independent firms and public bodies were involved in Italy's provision of water services at the end of 2013. Of these 1,235 operators, 75 % (n. 931) are municipalities or other public bodies (such as consortia of local governments or mountain communities) that provide one or more water services directly "in house." As can be seen in Table 2.2, the great majority of the local governments that have chosen to provide services directly (around 79 %) are located in the north of Italy, mainly in Lombardia and Trentino Alto Adige. In some regions (i.e., Basilicata, Friuli, Puglia, Sardegna, Umbria, and Veneto), no municipality or public body is involved in the provision of water services. In two regions (Molise and Valle d'Aosta), water services are provided only by municipalities or some other public body, with no water utilities involved in the industry (see Table 2.2). Moreover, only 232 municipalities or other public bodies

Table 2.1 Size of Italian water utilities

	Population served			Number of municipalities served		
	Water	Sewerage	Wastewater treatment	Water	Sewerage	Wastewater treatment
Average	124,224	116,046	138,240	12	12	16
Max	4,060,595	3,981,387	3,972,744	283	286	288
Min	31	23	79	1	1	1
Coverage of the sample (%)	55	46	43.10	38	34.70	35.40

Source (AEEG 2013)

2.3 An Overview of the Italian Water Industry

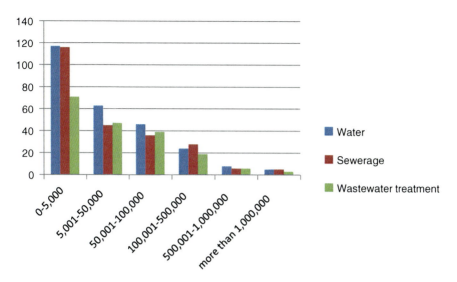

Fig. 2.2 Number of inhabitants served by utilities

Table 2.2 Geographical localization of public bodies and water utilities providing water services in Italy

Region	Area	Public bodies		Water utilities		Total	
		No.	%	No.	%	No.	%
ABRUZZO	South	9	1	7	2	16	1
BASILICATA	South	0	0	1	0	1	0
CALABRIA	South	2	0	3	1	5	0
CAMPANIA	South	49	5	13	4	62	5
EMILIA	North	3	0	8	3	11	1
FRIULI	North	0	0	9	3	9	1
LAZIO	Centre	42	5	9	3	51	4
LIGURIA	North	23	2	18	6	41	3
LOMBARDIA	North	321	34	82	27	403	33
MARCHE	Centre	17	2	12	4	29	2
MOLISE	South	51	5	0	0	51	4
PIEMONTE	North	28	3	31	10	59	5
PUGLIA	South	0	0	1	0	1	0
SARDEGNA	South	0	0	2	1	2	0
SICILIA	South	29	3	48	16	77	6
TOSCANA	Centre	1	0	10	3	11	1
TRENTINO	North	293	31	27	9	320	26
UMBRIA	Centre	0	0	3	1	3	0
VALLE D'AOSTA	North	63	7	0	0	63	5
VENETO	North	0	0	20	7	20	2
TOTAL		931	100	304	100	1235	100

Table 2.3 The specific type of services provided by 25 mono utilities

	Number of water utilities
Only collection	0
Only potabilization	1
Only adduction/transportation	0
Only wholesale	3
Only distribution of water for civil use	5
Only sewerage	1
Only wastewater	15
Total	25

provide all of the five main water services (i.e., the collection, transportation, and distribution of water for civil use and sewerage and wastewater treatment), while the others provide only one service or some (mainly sewerage and distribution).

Thus, only 304 of the 1,235 operators are independent firms (water utilities) that were established on average in 1991, so they are on average 23 years old, with a maximum of a firm that was established in 1852 (Società Acque Potabili, located in Turin).

Only 160 Italian water utilities provide at the same time the services of collection, transportation, and distribution of water for civil use, sewerage, and wastewater treatment. These utilities are located mainly in the north of Italy (64 and 27 % only in Lombardia). These data highlight a complex scenario, where there are regions (Basilicata, Puglia, and Sardegna) that have only one or two water utilities that manage the water services for the entire regional area, and regions (such as Lombardia, Trentino, and Sicilia) with numerous different operators.

Analyzing financial statements and websites, we find that many Italian water utilities provide only the water services (mono-utilities) and are not involved in other industries such as electricity, gas, or municipal waste management. Actually, 202 utilities are not involved in other businesses. In particular, 25 mono utilities provide only one service (see Table 2.3), while 108 firms are the mono utilities that provide at the same time all the main water services (collection, adduction/transportation, distribution of water for civil use, sewerage, and wastewater treatment).

The complexity of the water utilities' vertical integrations and diversification strategies makes it difficult to compare firms' performance and efficiency and reflects the complexity of the endogenous and environmental factors affecting decision makers' definitions of the best organizational structure for the water industry.

Using the AIDA database, we collect information about the number of employees, ownership type, and number of shareholders for each of the 304 utilities for 2012. We find that Italian water utilities had more than 43,700 employees, with an average of around 160 employees each, and a maximum of more than 6,500 employees in Hera Spa, the biggest Italian multi-utility. The mono-utility with the most employees was Acea Ato 2, serving the Roma area, followed by Abbanoa, which provides water services to almost all of Sardinia (both with around 1,400 employees). Thus, the water sector is very important for the Italian economy in

2.3 An Overview of the Italian Water Industry

Table 2.4 Clusters of firms on the basis of ownership type

Ownership type	Number of firms	% of firms	Average number of shareholders	Min number of shareholders	Max number of shareholders
Publicly-owned	162	53	29.67	1	343
Mixed-ownership	64	21	19.79	1	128
Privately-owned	78	26	29.90	1	583
Total	304	100	27.51	1	583

terms of employment; Romano and Guerrini (2014) show that Italian publicly owned water utilities have significantly more employees than the others do.

In addition, most of the 304 utilities (53 %) are public firms (whose shareholders are municipalities or other public bodies; see Table 2.4); 26 % are totally private firms, and the remaining 21 % are mixed-ownership firms with both public and private shareholders.

These 304 firms (excluding the 15 private partnerships and sole proprietorships, 13 co-ops, and 3 listed companies) have an average of 27 shareholders, with a minimum of one sole shareholder and a maximum of 583. The average number of shareholders is higher in private firms, although when excluding the firm with the most shareholders, the average is only 18.6, the lowest among the three clusters.

Moreover, 50 firms have only one shareholder, 33 of which are public; 90 firms (around 30 %) have no more than three shareholders, and only 13 have more than 100.

References

AEEG, Autorità per l'Energia Eletterica, il Gas e il Sistema Idrico (2013) Relazione annuale sullo stato dei servizi e sull'attività svolta. http://www.autorita.energia.it/allegati/relaz_ann/13/RAVolumeI_2013.pdf
Carrozza C (2011) Italian water services reform from 1994 to 2008: decisional rounds and local modes of governance. Water Policy 13(6):751–768
Co.N.Vi.Ri (2009) Rapporto annuale al parlamento sullo stato delle risorse idriche. Roma
Co.N.Vi.Ri (2011) Rapporto annuale al parlamento sullo stato delle risorse idriche. Roma
Danesi L, Passarelli M, Peruzzi P (2007) Water services reform in Italy: its impacts on regulation, investment and affordability. Water Policy 9(1):33–54
D'Angelis E, Irace A (2011) Il Valore Dell'Acqua. Dalai Editore, Milano
Guerrini A, Romano G, Campedelli B (2011) Factors affecting the performance of water utility companies. Int J Public Sector Manag 24(6):543–566
Guerrini A, Romano G (2013) The process of tariff setting in an unstable legal framework: an Italian case study. Utilities Policy 24:78–85
Marques R (2010) Regulation of water and wastewater services. An international comparison. IWA Publishing, London
Romano G, Guerrini A (2014) The effects of ownership, board size and board composition on the performance of Italian water utilities, working paper
Testa F (2010) A proposito di acqua e servizi pubblici locali. Manag delle Utilities 1:97–98

Chapter 3
The Determinants of Water Utilities Performance

3.1 The Effects of Ownership and Political Connections on the Performance of Water Utility Companies: An Overview

Over the last 25 years, the governance of public services has undergone important reforms in many countries. During the 1990s, efforts to reform the corporate entities established to pursue public policy and commercial objectives wholly owned either by the state or local governments (state-owned enterprises, or SOEs) were aimed at promoting privatization even if, for both political and economic reasons, the state remained a major owner of productive assets in many economies (Menozzi et al. 2011).

Some authors argue that SOEs perform less efficiently and less profitably than private firms (Shleifer and Vishny 1994; Boycko et al. 1996) and that ownership (OWN), together with competition, is important in promoting efficiency (Boardman and Vining 1989; Bozec and Dia 2007). Privatization is thus considered an appropriate way to achieve significant improvements in SOE performance (Megginson et al. 1994; Shleifer 1998; Dinc and Gupta 2011; Dewenter and Malatesta 2001; Arocena and Oliveros 2012).

Water services have accordingly been privatized in several countries, notwithstanding the conflicts between the profit-seeking behavior of private partners and the public objectives of the water services (Hall 2001). The UK, France, Portugal, Spain, and Italy have all pursued privatization, with mixed results (see Abbott and Cohen 2009; Berg and Marques 2011). Privatization in the water industry has had conflicting consequences on efficiency and profitability (Bakker 2003; García-Sánchez 2006; Lobina and Hall 2007; Carrozza 2011) as well as on investment and financial structures (Shaoul 1997; Vinnari and Hukka 2007; Romano et al. 2013). Two studies demonstrate that private utilities outperformed public companies in consuming certain production factors such as labor (Picazo-Tadeo et al. 2009a, b). One group of scholars reported that the OWN structure did not influence performance (Byrnes et al. 1986; García Sánchez 2006; Kirkpatrick et al. 2006; Seroa da Motta and Moreira 2006).

The water industry is capital-intensive. A number of scholars (e.g., Idelovitch and Klas 1997; Yamout and Jamali 2007) and international organizations (e.g., OECD and the World Bank) support water industry privatization, arguing that the funding of water and wastewater utilities exceeds public sector capabilities and that privatization represents a promising solution to the water supply problem. Recently, however, Hall and Lobina (2012) have argued that public firms fund investments in the water sector more effectively both in developed and developing countries. Hall and Lobina point to three main advantages of public finance: first, the state pays lower interest rates than private investors; second, the state grants all citizens access to water services even if they cannot afford to pay the whole cost; and finally, the health benefits of water and sanitation networks are social rather than private gains. Moreover, private investors have less incentive to invest in the water industry since massive sunk costs represent a significant share of total costs (Ménard and Saussier 2000), and the payback period is prolonged. Private investors are therefore conscious that investments can be recovered only after many years (Idelovitch and Klas 1997; Massarutto et al. 2008). Hassanein and Khalifa (2007) highlight how the water industry is incapable of effectively attracting private participants because the status of the water system is unknown, as most of the assets of water and wastewater utilities are underground. Moreover, private firms take into account the losses associated with inadequate systems, such as revenue collection and water leaks. As shown by Massarutto et al. (2008), the cost of capital has a decisive impact on water utilities' investment decisions. They argue that, on the one hand, public funding is cheap but scarce, as well as untimely and even potentially harmful (since it may encourage inefficient investment choices); private funding is potentially unlimited, on the other hand, and inspires efficient behaviors but is unduly costly and may lead to tariffs above the real economic cost. For these reasons, Massarutto et al. (2008) conclude that delegating all responsibilities and risks to private operators may lead to unsustainable tariff increases when major investments are needed.

In Italy, there is much debate on the privatization of water firms, which were originally owned by local governments. Italian municipalities have historically provided public services directly through public administrations; during the 1990s, however, legislation transformed many municipal utilities into corporations regulated by private law (see Chap. 2). Thus, the Italian water industry has been transformed over the last 20 years through extensive legislative reforms designed with the aim, among others, to end the in-house supply of water and wastewater services by outsourcing them to independent public, mixed, or private firms.

Studying the Italian context, Guerrini et al. (2011) find that private utilities are, as expected, more oriented toward profit, since their financial ratios, such as return on sales (ROS), are twice those of public companies, and financial leverage is used more intensively. Romano and Guerrini (2011) point out that public Italian water utilities have the highest efficiency scores, since they purchase and employ inputs more efficiently than do mixed-OWN firms. Similarly, Cruz et al. (2012)

show that, in both Italy and Portugal, water utilities with public OWN are more efficient than mixed and private ones, demonstrating that the reforms favoring private sector participation in both countries were not necessarily successful. Interestingly, mixed companies appear to be more efficient than totally private companies, contradicting the literature, in which mixed-OWN is often seen as the worst scenario (Eckel and Vining 1985; Boardman and Vining 1989; Cruz and Marques 2012). Romano et al. (2013) find that public Italian water utilities have healthier financial structures than do mixed-OWN firms, with higher solvency and independence ratios.

As highlighted in Chap. 2, the great majority of Italian water industry operators were, at the end of 2013, still municipalities or other public bodies. Among 1,235 operators, only 304 were independent firms, most of them totally public (with shareholders that are municipalities, other local government, or another kind of public body). Thus, mixed and private utilities now coexist with the majority of firms that are totally public.

Moreover, privatization without a transfer of control seems unlikely to favor efficiency or profitability unless firms' choices are shielded from the influence of politicians and bureaucrats (Li and Xu 2004; Gupta 2005; Shleifer and Vishny 1994. The composition of the board of directors is a central performance factor, since the board defines the firm's corporate and business strategy and has an important advisory role (Adams et al. 2010; Agrawal and Knoeber 2001). Thus, if privatized firms and their boards do not gain complete independence from national and local government influence, they are likely to face conflicting objectives, and politically connected firms will probably exhibit poorer accounting performance than their nonconnected counterparts will (Fan et al. 2007; Sørensen 2007; Boubakri et al. 2008). However, Agrawal and Knoeber (2001) argue that politically experienced directors aid their firm with their knowledge of government procedures and their ability to predict government actions; they can also help forestall government actions inimical to the firm. The presence of politicians on a board has been investigated, the evidence showing a positive effect of political connectedness on firm value and performance (Faccio 2006; Goldman et al. 2009; Niessen and Ruenzi 2010). Faccio (2010) shows that politically connected corporations have, on average, higher leverage, enjoy marginally lower taxation, and display much greater market power; they also have lower ROA and market valuation than their peers, however. Similarly, Menozzi et al. (2011) show that, for local Italian public utilities (operating not only in the water industry but also in the gas and electricity sectors), politically connected directors exert a positive and significant effect on employment but have a negative impact on profitability. Recently, Romano and Guerrini (2014) have shown that boards of Italian water utilities in 2011 were dominated by politically connected directors who negatively affected the firms' financial structures without influencing their economic performance. The authors find that private or mixed-OWN utilities show higher profitability than do totally public firms, though the latter are less debt-dependent and have more employees. Thus, the effects of public OWN and political connectedness have been shown to be positive on employment and negative on profitability.

3.2 Boosting Efficiency Through Economies of Scale, Scope, and Population Density: Evidence from Prior Studies

Many empirical studies conducted worldwide have addressed the relationship between the performance of water utilities and their size and diversification, as well as the possible existence of economies of scale, scope, and density.

Economies of scale arise when a unit increase in output results from a less than proportional increase in input. Economies of scope occur when an entity's unit average cost to produce two or more products or services is lower than that when they are produced by separate entities. The water industry has two types of economies of density: (1) output or production density, the extent of the change in costs when the total volume of water produced or wastewater treated increases while the number of customers and network length remain constant, and (2) customer density (CD), the quantum of change in costs when the number of customers increases while constant network length remains constant (Nauges and Van den Berg 2008).

About this issue, Saal et al. (2013) reviewed the theoretical definitions of the measures of economies of scale and scope applied in the literature, discussed the characteristics of the cost functions underlying the empirical estimation of these measures, and reviewed the literature on economies of scale and scope. In addition, Guerrini et al. (2013) summarized the findings of the most relevant research papers on the effects of scale, scope, and density on the performance of water utility companies. However, the results lack consensus. Most studies confirmed the presence of economies of scale in the water industry (e.g., Carvalho and Marques 2014; Guerrini et al. 2011; Shih et al. 2006), but several others found diseconomies of scale in various countries (e.g., Aida et al. 1998; Alsharif et al. 2008; Antonioli and Filippini 2001; Bhattacharyya et al. 1995; Ford and Warford 1969; Mizutani and Urakami 2001; Saal and Parker 2000; Saal et al. 2007).

By contrast, as highlighted by Abbott and Cohen (2010), there is a consensus that economies of scale do exist for wastewater activities, although there is no clarity on the timeframe of its availability. However, this consensus might exist only because the wastewater industry has received less research attention than others industries have, with most studies focusing on firms that conduct water supply as well as wastewater activities (e.g., Ashton 2000; Romano and Guerrini 2011).

Many studies on economies of scale propose that only small- and medium-sized firms can improve efficiency through expansion and that big firms do not always benefit through expansion and sometimes even suffer diseconomies (De Witte and Marques 2011; Filippini et al. 2008; Kim and Clark 1988; Marques and De Witte 2011; Martins et al. 2006; Torres and Morrison-Paul 2006). In addition, researchers have not agreed on an "optimal scale" (see Guerrini et al. 2013), which appears to vary considerably among countries (Saal et al. 2013).

Consensus is also lacking on the existence of economies of scope (González-Gómez and García-Rubio 2008; Guerrini et al. 2011). The vast majority of the research pertains to vertical integration in the water industry value chain (i.e., production and distribution, water and wastewater, and water and sewage). Vertically integrated water utilities are the most common, and hence, most researched kind in many countries because most water supply services are managed locally, ensuring that production plants and distribution networks are in close proximity. Multiple water suppliers operating in the same distribution network may create problems involving the compatibility of water treatments, the origin of water in the network, or liability for sanitary problems (García et al. 2007). Economies of vertical integration have been found to exist when a single firm is able to produce the complementary products of an industry's successive production stages more efficiently than several different firms can (García et al. 2007).

While the evidence shows that economies of scope exist for water production and distribution (Saal et al. 2013), the results of studies on the joint provision of water and wastewater services differ. Studies focusing on the UK and Portugal (see Guerrini et al. 2013; Saal et al. 2013) offer contrasting results. Lynk (1993) and Hunt and Lynk (1995) find economies of scope, while more recent studies (Saal and Parker 2000; Stone and Webster Consultants for OFWAT 2004) find diseconomies of scope. Studying Portugal, Martins et al. (2006) and Carvalho and Marques (2013) find economies of scope, while Correia and Marques (2011) and Marques and De Witte (2011) find diseconomies of scope. Moreover, Carvalho and Marques (2011) show that the simultaneous provision of water supply and wastewater services (rather than the sole provision of water supply services) hinders performance. However, they observed economies of scope in Portugal, where a positive influence resulted from the joint provision of water supply, wastewater, and urban waste services. On the other hand, using Wisconsin data, García et al. (2007) show that separating production and distribution stages might lead to cost savings, although not for the smallest services.

The results for the wastewater treatment sector are not convergent. Knapp (1978), studying the UK, found economies of scale of up to 16.6 million m^3/year. Similarly, Rossi et al. (1979) confirmed the possibility of achieving economies of scale by increasing sizes. More recently, after studying the Danish sector, Guerrini et al. (2014) found that strategies aiming to extend the area served by wastewater utilities (such as covering new areas or merging with other companies) do not yield cost savings. Concerning Italy, Fraquelli and Giandrone (2003) found economies of scope from vertical integration and strong economies of scale for smaller structures.

Finally, the literature has addressed economies of density (Caves et al. 1981). In the water industry, economies of density exist when unit costs decrease with greater population density or with an increase in the water provided per kilometer of mains, because the costs of the infrastructure required to provide the service is lowered. Thus, water utilities have significant economies of both customer and output density (see Guerrini et al. 2013). Since differences in population density are likely to influence utility costs and vertical integration economies (Saal et al. 2013), further research is needed to fully understand this issue.

The evidence concerning economies of scale in the Italian water industry is conflicting. Fabbri and Fraquelli (2000) find large economies of scale for firms that deliver a minimum of around 350,000 m^3 and diseconomies at the maximum point of 393,960,000 m^3. Focusing on multiutilities, Fraquelli et al. (2004) find economies of scale only for firms with output levels lower than those that characterize a median firm; those bigger than the median experienced neither economies nor diseconomies of scale. By contrast, Romano and Guerrini (2011) demonstrate that economies of scale in Italy also apply to firms in the medium cluster (those with more than 50,000 customers). Guerrini et al. (2013) demonstrate that diseconomies appear for DMUs that collect less than 60 million € in revenues; beyond this threshold, each firm records the maximum VRSTE. The biggest players in this sector (with revenues of 100–400 million €), such as Hera, AQP, AcegasAps, and Metropolitana Milanese, achieved optimal efficiency levels, probably because of their excellent pipeline capacity, skilled staff, and better purchasing power for strategic inputs (i.e., electricity and services).

Concerning economies of scope, Guerrini et al. (2011) have analyzed the effects of diversification in water-related industries, such as the electricity, gas, and urban waste industries. Economies of scope characterize the water sector and depend on factors other than labor costs, such as energy costs, overhead, and discretionary costs. In addition, Italian multiutilities incur higher labor costs per capita than do monoutilities, likely because they are more complex and require more highly skilled managers and employees. Similarly, in an analysis of 90 Italian utilities operating in the gas, water, and electricity sectors from 1994 to 1996, Fraquelli et al. (2004) find significant economies of scope among multiutilities with output levels lower than the median, highlighting how small, specialized firms might benefit from cost reductions by transforming into multiutilities providing more than one service, such as gas, water, and electricity concurrently. The highest cost advantage stemmed from the joint provision of water and gas.

Finally, examining economies of density, Fabbri and Fraquelli (2000) find that greater CD leads to cost savings, an effect confirmed by Antonioli and Filippini (2001) and Guerrini et al. (2013).

The conflicting results seen for economies of scale in the Italian water sector and the scarce evidence for economies of CD lead us to investigate these important issues more thoroughly.

3.3 Data Collection and Research Method

3.3.1 Data Collection and Description

Starting with the 108 companies that provide all the main water services (i.e., collection, adduction/transportation, distribution of water for civil use, sewerage, and wastewater treatment) using the Bureau Van Dijk AIDA database and information available from corporate websites, we collected data on the populations these utilities served, their main lengths, and their financial statements. In contrast to

3.3 Data Collection and Research Method

previous empirical research (Guerrini et al. 2013), this study included only monoutilities, thus eliminating from its statistical analysis the effect of differentiated operations and strategies, which could severely affect firm performance (Guerrini et al. 2013).

Data on the length of the mains and the number of inhabitants served were generally available from corporate websites or financial statements; otherwise, we solicited this information directly from company technical staff. The financial statements were obtained through the Bureau Van Dijk AIDA database, which gave us data on revenues, value of production, depreciation, amortization and interest paid, staff costs, and other operative costs (e.g., services, maintenance, materials). Finally, the number of employees was collected.

We were able to obtain complete information on about 98 out of the 108 companies that provide water services to approximately 57 % of Italians. Our central data set thus accounts for most of Italy's water industry. Data on main lengths and population served were not found for 10 firms.

We observed a period of 5 years (2008–2012) during which the water sector was conditioned by intense reform (see Chap. 2): (1) law 135/2009 imposed a privatization process for public water utilities; (2) this provision was abrogated through the 2011 referendum that also delayed the compulsory return on invested capital, set at 7 % of invested capital by the tariff method (the so-called "Normalized method").

In light of these changes, it would be interesting to examine their effects on efficiency and determine which types of firm have been conditioned the most. Finding all five financial statements (2008–2012) was impossible for 11 of the 98 firms: eight did not disclose their financial figures for one of the 5 years monitored, while three were constituted after 2008.

During the 5 years under study, the 98 selected water utilities generate a turnover of about 20.5 billion €. The annual mean production value (PV) is 43.6 million, while the staff employed is, on average, 175 units. These data reveal the size of the monoutilities providing water services in Italy. According to EU parameters, firms with more than 50 million € in sales are "large," those that earn between 50 and 10 million € are "medium," and those with less than 10 million € in revenues are "small" utilities. Nevertheless, observations are highly dispersed: some small Italian firms without internal staff provide water services using the employees of the municipality that own the firm, which coexist with large corporations employing thousands of workers and listed on a stock exchange (Acque Potabili). This difference is also seen when the characteristics of the served areas are considered: the population served varies between 4 million and 4,400 inhabitants; similarly, main lengths reach from 22,500 km for larger firms to no more than 100 km for small ones. Tables 3.1 and 3.2 provide a composite picture of the sector, featuring wide differences in population density: some water utilities operate with a network density (i.e., population served to main length) of more than 1.000 inhab./km of mains, while others serve only a few dozen citizens with 1 km of mains.

We divided this data set into groups according to four criteria. First, we categorized companies into large, medium, and small utilities according to the

Table 3.1 Brief descriptive statistics of "financial statement" items

	Production value	Staff cost	Capex	Amort.	Interest paid	Oopex	Staff
Mean	43,599,734	9,541,787	6,406,934	5,326,115	1,080,819	26,911,547	175
Max	552,306,126	110,314,510	121,279,925	91,029,831	30,250,094	285,397,197	2,113
Min	269,202	–	682	517	–	56,434	–
St. dev.	78,376,006	16,798,214	14,064,653	11,178,601	3,326,919	44,251,446	352

Table 3.2 Brief descriptive statistics of environmental and operational variables

	Population served	Main length (km)	Density
Mean	425,248	2,695	158
Max	4,000,000	22,500	1,124
Min	4,420	60	9
St. dev.	619,237	3,117	172

abovementioned EU parameters. Then, firms were grouped by their localization (LOC) on the Italian peninsula, divided into northern, central, and southern firms.

Two clusters were created according to OWN structure: totally public firms (generally controlled by a network of municipalities) and mixed-OWN or fully private firms. This was done because only a few (nine) of the utilities are fully private, as the vast majority of private firms are involved in only one water service or a few (such as distribution). Finally, measuring the ratio of the population served to kilometers of mains, we identified three approximately equally sized groups, based on their customer densities: high density (HD; ≥ 153 inhab./km), medium density (MD; 153 inhab./km< >86 inhab./km), and low density (LD; ≤ 86 inhab./km).

Table 3.3 provides an overview of the clusters, along with their descriptive statistics. The clusters differ substantially in their representation of the Italian context. Some firms are 30 times smaller than others when measured by PV, 20 times smaller in terms of average population served, and 10 times smaller in terms of km of mains length. As expected, mixed and private firms are larger than public ones in terms of PV; however, the latter have higher average staff costs, implying that public control maximizes the number of workers to the detriment of efficiency. This evidence will be discussed in depth during the statistical analysis.

The geographical distribution of water utilities differs among the north, center, and south of Italy. The northern one is characterized by a high number of utilities (273), 45 % of which have a PV lower than 10 million €.

The center of Italy mainly features large and mixed-OWN firms; in effect, this area has progressively developed a process of utilities aggregation to reach economies of scale, in the spirit of the Galli Law.

The southern firms are, on average, larger than the northern and are more labor-intensive than are the other two clusters. This could reveal inefficiencies, as will be discussed in the following analysis.

Finally, most firms operating in densely populated areas are larger than others, and 40 % of them are run by public–private partnerships. This interesting result,

3.3 Data Collection and Research Method

Table 3.3 Average value for clusters of firms

Average value	Production value	Staff cost	Capex	Oopex	Population served	Mains length (km)	Density
Size							
Large	153,315,290	32,002,679	24,867,818	88,760,503	1,193,843	6,755	192
Medium	25,910,591	5,557,311	3,741,199	16,690,185	224,681	1,804	188
Small	4,890,710	1,195,925	610,895	2,948,628	58,803	613	141
Localization							
North	28,970,714	6,432,351	3,446,420	18,406,696	223,107	1,513	161
Centre	74,892,983	13,948,065	15,210,404	38,290,034	525,054	4,019	144
South	53,852,163	12,103,396	8,339,680	34,261,104	540,620	3,070	218
Ownership							
Public	40,910,555	9,307,314	5,504,531	25,879,230	346,285	2,354	170
Mixed and private	47,443,974	8,904,671	8,950,010	25,755,441	369,187	2,319	175
Cluster density							
High density	74,946,447	16,006,478	11,064,132	45,051,955	652,082	2,560	334
Medium density	34,183,922	7,128,682	5,883,439	20,023,797	271,352	2,248	121
Low density	20,543,368	4,319,442	3,280,916	12,342,286	138,896	2,216	60

according to Ménard and Saussier (2000), implies that those areas requiring much larger investments (those with scarce water sources or with low population densities) attracted the direct management of public bodies to avoid opportunistic behavior by private operators.

3.3.2 DEA Analysis

To detect economies of scale and density in the Italian water industry and evaluate the impact of OWN and LOC on efficiency, we applied a two-stage method, based on DEA and regression analysis. Unlike previous research (Cubbin and Tzanidakis 1998), we did not use regression analysis solely as a control method to confirm the DEA results. Instead, after creating a ranking based on the DEA scores, we applied a regression model to determine the influences of the four independent variables (i.e., PV, CD, OWN, LOC). This two-stage method has been applied to study the water sector (Estache and Kouassi 2002; Anwandter and Ozuna 2002; Kirkpatrick et al. 2006; García-Sànchez 2006; Renzetti and Dupont 2009; Guerrini et al. 2013).

As a nonparametric technique, the DEA can determine a frontier and calculate an efficiency ratio for each decision-making unit (DMU). Through a linear programming approach, the DEA identifies an efficient virtual producer for each unit; the efficiency ratio is the distance separating the virtual from the real unit. Charnes et al. (1978) use this linear programming method to build a production frontier, in which DMUs can linearly scale inputs and outputs without any variation in efficiency. However, this assumption is valid only for a limited range of production, when all units operate on an optimal scale. Thus, Banker et al. (1984) remove the constant return to scale (CRS) assumption and instead determine a scale effect (SE) and a pure technical efficiency (VRSTE), which, combined, yield a global efficiency index (CRSTE). The VRSTE measures a company's real capability to purchase, mix, and consume inputs, and its SE indicates the effectiveness of the decision to operate at a certain production scale. To evaluate the SE, we must consider the distance between the variable return scale (VRS) frontier and the CRS frontier for each DMU. In line with most DEA research (Berg and Marques 2011), we opt for the VRS assumption and thereby highlight the real determinants of global efficiency in water utilities by distinguishing pure from scale efficiency.

If efficiency is the capability to reduce the consumption of inputs at a given level of output, we must choose the measures used as inputs and outputs in the DEA model carefully. According to two analyses (Berg and Marques 2011; De Witte and Marques 2010), the most frequently adopted inputs are staff cost, operational expenditures, energy, and mains length.

The leading output measures are the distributed water volume and the number of customers. Consistent with this evidence and the data available, we consider four inputs, the sum of amortization, depreciation, and interest paid, staff costs, other operating costs, and the length of the mains, and two outputs, population served and PV.

To solve the chosen linear programming model we used DEAP Version 2.1 (Coelli 1996), a freely downloadable software for efficiency analysis developed by the Centre for Efficiency and Productivity Analysis (CEPA). This software allows users to define their own linear programming model by choosing the kind of return scale (constant or variable), orientation toward either input or output, and the number of stages needed to solve the problem. We have already addressed the choice of return scale assumption. We discuss the other two aspects below.

Input-oriented models define an efficiency improvement as a proportional reduction in input consumption and outputs, whereas output-oriented models view efficiency as an increase in output production given a certain amount of input. Scholars use the output orientation model when the DMUs being observed have a certain amount of resources and must maximize outputs; if DMUs need to produce a fixed level of output but aim to reduce their input consumption, an input-oriented model is more appropriate. For water utilities, outputs (measured by cubic meter of water delivered or inhabitants served) remain fairly constant over time, but inputs depend on price fluctuations and internal efficiency. Therefore, most of the relevant research uses input-oriented models (Berg and Marques 2011; De Witte and Marques 2010).

We adopted the following linear programming model, with the assumption of a VRS and input orientation:

3.3 Data Collection and Research Method

$$\text{Min } \Phi$$

$$\sum_j \lambda_j x_{jm} \leq \Phi x_{j_0 m}; \quad m = 1, 2, \ldots, M$$

$$\sum_j \lambda_j x_{jn} \geq y_{j_0 m}; \quad n = 1, 2, \ldots, N$$

$$\lambda_j \geq 0 \ \& \ \sum_j \lambda_j = 1; \quad j = 1, 2, \ldots, J$$

With DEAP 2.1, we can choose between a one-or multistage model. The efficient projected points determined by a one-stage DEA model (Charnes et al. 1978) may not comply with the criterion of Pareto optimality, in which case they should not be classified as efficient points, a problem due to the input/output slacks that arise when it is still possible to increase outputs or reduce inputs beyond an efficient projected point on the frontier. Following Coelli (1998), therefore, we adopted a multistage linear programming model that can set aside slacks and give a Pareto-optimal set of projected points.

3.3.3 Statistical Analysis

The final part of this section provides a description of the statistical analysis applied to the DEA scores (CRSTE, VRSTE, SE).

Existing DEA studies seek to group DMUs using exogenous and operational variables, such as geographical LOC or size, to identify influences on efficiency (Brockett and Golany 1996; Anwandter and Ozuna 2002; Romano and Guerrini 2011; García-Sánchez 2006). To achieve our similar objective, we conducted a statistical analysis comparing the means, medians, and variances of the DEA scores for the created clusters. When differences are statistically significant, the variable used to group firms is a relevant determinant of performance.

We thus applied median and t-tests to reveal the differences between the two clusters created on the basis of OWN (i.e., public and mixed-private utilities); a Bartlett's test indicated the differences across groupings based on size (large, medium, or small), density (HD, MD, or LD), and LOC (north, center, or south). Nonparametric rank statistics, such as the Mann–Whitney test, are particularly appropriate for testing DEA outcomes because the distribution of their efficiency scores is generally unknown (Brockett and Golany 1996). We applied a Mann–Whitney test to verify the differences between public and mixed-private firms.

Next, we used a regression model to verify the findings of these tests and explore the causal relationships further. The model related each DEA score to four independent variables:

- Production value, a continuous variable measuring firm size to detect the presence of scale economies.
- Customer density, indicating the presence of economies of density in the Italian water industry, measured by the ratio of population served to kilometers of main length.

- Localization, a dummy variable reflecting the geographical area where the water utilities operate (i.e., north, center, or south).
- Ownership, a dummy variable reflecting the firm's OWN (i.e., public or mixed-private).

We ran the model three times, once for each DEA score (CRSTE, VRSTE, SE), as follows:

$$\text{DEA SCORES} = \beta 0 + \beta 1 PV + \beta 2 CD + \beta 3 LOC + \beta 4 OWN + \varepsilon.$$

We chose a Tobit regression function because of its ability to describe the relationship between a non-negative dependent variable and the independent variables. Scholars frequently associate Tobit functions with DEA models when studying performance across several industries because the dependent variable value, measured by DEA scores, is restricted between 0 and 1 (Aly et al. 1990; Chirkos and Sears 1994; Dietsch and Weill 1999; Ray 1991; Sexton et al. 1994; Stanton 2002). However, this two-stage approach has been criticized (Simar and Wilson 2004, 2007) for failing to account for serial correlation in DEA scores. Because DEA scores may be biased and as the environmental variables correlate with output and input variables, bootstrapping techniques can more clearly reveal the impact of exogenous and operational variables on efficiency scores (Peda et al. 2013).

Despite the limits of our chosen two-stage method, it offers an appropriate means of answering our research question and has been widely used (Tupper and Resende 2004; García-Sánchez 2006; Renzetti and Dupont 2009) because of its superior effectiveness compared with alternative approaches, such as ordinary least squares, the Papke–Wooldridge Method, and the unit inflated beta model (Hoff 2007).

Finally, since we were working with observations of multiple phenomena obtained over five time periods for the same firms, we used a panel data Tobit regression, which takes account of the correlations among observations for each utility during the years analyzed.

3.4 Results and Discussion

This section reports the results of the two-stage DEA model. Table 3.4 presents the descriptive statistics of the DEA scores. The high mean values and distribution of the DEA scores are not widely dispersed: each firm is ranked over 0.60. These results are far above those obtained in prior research on the Italian water sector. Romano and Guerrini (2011) collected financial statements and other technical data for 43 monoutilities in 2007 and obtained 0.14 as the global efficiency score (CRSTE) and 0.37 for technical pure efficiency (VRSTE). Guerrini et al. (2013) examined the same kind of data for 64 mono and multiutilities in 2008: they found significant efficiency improvements, with an average CRSTE of 0.78 and a VRSTE of 0.83, though the average scores were lower than those detected in the current research (which obtained a CRSTE of 0.88 and a VRSTE of 0.90). The difference between the 2007 and 2008 results could be explained in terms of

3.4 Results and Discussion

Table 3.4 Brief descriptive statistics of DEA scores

	CRSTE	VRSTE	Scale
Mean	0.88	0.90	0.98
Max	1.00	1.00	1.00
Min	0.60	0.60	0.77
St. dev.	0.08	0.08	0.03

Table 3.5 DEA scores time series

Year	CRSTE	VRSTE	Scale
2008	0.876	0.896	0.979
2009	0.883	0.903	0.979
2010	0.883	0.903	0.978
2011	0.881	0.900	0.979
2012	0.884	0.901	0.981

economies of scope: the inclusion of multiutilities in the 2008 sample improved its average efficiency, since cost savings can be obtained through the provision of more than one public service. The progressive efficiency improvement recorded from 2008 to 2012 could thus be due to the capabilities and skills acquired by the firms during the 5 years observed (see Table 3.5).

Efficiency seems to vary among the clusters defined, though the gaps are not wide.

Public firms show lower DEA scores than do mixed and private firms, a difference confirmed by every test used (i.e., t-test, median test, and Mann–Whitney test) with a high degree of significance (1 % in six out of nine tests). These first results suggest that efficiency was not sufficiently stressed by public firms, which probably pay more attention to other aims (e.g., low tariffs, water savings, sustainability). The results are only partially confirmed by the regression model: only CRSTE is positively affected by a mixed-private OWN, while VRSTE and SCALE do not vary significantly, implying that mixed and private firms perform better than public ones but only in terms of global efficiency: when pure technical efficiency (the capability to purchase and consume input) is considered, OWN structure is not a significant variable.

Comparing the results shown in Table 3.6 to those in Romano and Guerrini (2011) reveal interesting differences: the prior empirical research suggests that public firms are more efficient, a finding that may have been negatively affected by the smaller number of observations (43) collected for that study; this project collected 473.

Examining the data on size, we note that the smallest and largest firms perform better than medium firms in terms of global and pure technical efficiency: small firms achieve the best CRSTE and SCALE scores, and the large utilities have the highest VRSTE score. Thus, economies of scale could affect those water utilities that collect more than 50 million in revenue. This result is confirmed by the regression model: the VRSTE is positively influenced by the PV. Therefore, the capability to purchase and consume input grows with turnover. This finding is quite robust (with a

Table 3.6 Testing the differences among clusters

	CRSTE	VRSTE	SCALE
Ownership			
Public	0.87	0.89	0.97
Mixed and private	0.89	0.91	0.98
T-test	0.000***	0.017**	0.002***
Median test	0.000***	0.000***	0.001***
Mann–Whitney	0.000***	0.010**	0.087*
Size			
Large	0.88	0.92	0.96
Medium	0.87	0.89	0.98
Small	0.89	0.91	0.98
Bartlett's test	0.469	0.202	0.017**
Median test	0.040**	0.020**	0.000***
Localization			
North	0.89	0.91	0.98
Center	0.88	0.89	0.99
South	0.86	0.89	0.97
Bartlett's test	0.000***	0.000***	0.000***
Median test	0.030**	0.822	0.018**
Cluster density			
High density	0.91	0.93	0.97
Medium density	0.87	0.89	0.98
Low density	0.86	0.88	0.98
Bartlett's test	0.173	0.926	0.000***
Median test	0.000***	0.000***	0.17

***, **, and * indicate 1, 5, and 10 % significance levels, respectively

significance lower than 1 %) and thus should be a real feature of the Italian water sector. Contradicting the results of the parametric and nonparametric tests, Table 3.7 shows that global efficiency increases with turnover. This result for VRSTE was also obtained by a prior study (Guerrini et al. 2013), which demonstrates that diseconomies appear for DMUs that collect under 60 million € in revenues; beyond this threshold, each firm records a maximum VRSTE. Other research indicates that growth advantages accrue only to small firms, whereas similar strategies followed by large companies lead to diseconomies (Italy has been studied by Fraquelli and Giandrone 2003; see also Torres and Morrison-Paul 2006; Tynan and Kingdom 2005; Sauer 2005; Martins et al. 2006; Filippini et al. 2008; Marques and De Witte 2011).

Localizations on the Italian peninsula also play a key role in determining efficiency. The results shown in Tables 3.6 and 3.7 are convergent for CRSTE and VRSTE: northern firms are more efficient than are southern ones. This is so for two main reasons: (1) northern firms try harder to avoid wasting resources, and (2) the south has a more complex environment, characterized by water scarcity and older mains that suffer the highest rate of water loss in Europe (more than 50 % of it).

Table 3.7 The regression model

	CRSTE	VRSTE	Scale
Production value	0.000**	0.000***	0.000
Density	0.0002***	0.0002***	0.000***
Localization			
Center	−0.028	−0.040	0.001
South	−0.053***	−0.045**	−0.017**
Ownership			
Mixed-private	0.028**	0.017	0.007

***, **, and * indicate 1, 5, and 10 % significance levels, respectively

These findings are consistent with Guerrini et al. (2011), who have highlighted the lowest labor costs relative to PV of northern Italian water firms. Conversely, conflicting results appear in Romano and Guerrini (2011), where central-southern firms obtain the highest CRSTE scores; aggregating these firms into a single cluster may have counterweighed the inefficiency of the southern ones.

Finally, CD is the environmental variable that exerts the most relevant impact on efficiency: each DEA score is positively affected by density. Firms operating in small, densely populated areas such as cities achieve the lowest costs of delivering a cubic meter of water and often apply higher rates. This cost advantage depends on the ability to deliver a cubic meter of water with fewer resources (i.e., mains and electricity) and to limit water losses per customer. These findings are also consistent with the prior research (Fabbri and Fraquelli 2000; Tupper and Resende 2004; García-Sánchez 2006) and with the 2008 data (Guerrini et al. 2013). However, unlike Carvalho and Marques (2011), we find no threshold value for CD.

3.5 Conclusions

Though many studies have examined the environmental and operational variables affecting the efficiency of Italian water utilities, this study uses a larger dataset comprising 98 companies observed over 5 years, for a total of 473 observations.

Four variables widely used in empirical studies were chosen—OWN structure, firm size, geographical LOC, and CD—which are all at least somewhat controllable by the firms owners and managers, who can choose their OWN structure and select their operating region according to criteria such as surface extension, the number of inhabitants per square kilometer, and other physical and geographical characteristics. The only controllable variable for municipalities is public utility OWN; all others are given.

Our research findings are strongly convergent concerning CD. Cost advantages accrue to firms operating in regions with high population density because the presence of many customers per kilometer of mains reduces the costs of delivering a cubic meter of water through the lower unit costs of energy and infrastructure.

This variable has to be carefully considered by investors operating in the water sector and by policy makers and regulatory authorities when choosing contract arrangements and planning tariff models to suit specific areas.

Another milestone for water efficiency was found: firms need to grow to collect cost savings and thus increase profits. Growth strategies should be deployed by the shareholders and managers of water utilities, who need to find new areas to serve through mergers and acquisitions or partnerships. This would dramatically improve firms' economies of scale and bargaining power with suppliers, workers, and authorities, thus leading to better conditions for water services provision.

Finally, the overall evidence offered by previous studies suggests less robust and partially conflicting results for OWN structure and LOC. However, this research, based on a wider dataset, shows that the presence of a private shareholder should improve global efficiency, chiefly through the adoption of technological innovations, job training, well-defined procurement policies, and the development of an internal control system dedicated to achieving effective and efficient corporate processes. Moreover, the highest cost savings are achievable by firms localized in the north of Italy, probably due to its more favorable climate and geographical characteristics and better infrastructure, such as water mains and wastewater treatment plants.

References

Abbott M, Cohen B (2010) Industry Structure Issues in the Water and Wastewater sector in Australia. Econ Pap: J Appl Econ Policy 29(1):48–63

Abbott M, Cohen B (2009) Productivity and efficiency in the water industry. Utilities Policy 17:233–244

Adams RB, Hermailin BE, Weisbach MS (2010) The role of boards of directors in corporate governance: a conceptual framework and survey. J Econ Lit 48(1):58–107

Agrawal A, Knoeber CR (2001) Do some outside directors play a political role? J Law Econ 44(1):179–198

Aida K, Cooper WW, Pastor JT, Sueyoshi T (1998) Evaluating water supply services in Japan with RAM: a range-adjusted measure of inefficiency. OMEGA. Int J Manage Sci 26(2):207–232

Alsharif K, Feroz EH, Klemer A, Raab R (2008) Governance of water supply systems in the Palestinian territories: a data envelopment analysis approach to the management of water resources. J Environ Manage 87:80–94

Aly HY, Grabowski R, Pasurka C, Rangan N (1990) Technical, scale and allocative efficiencies in US banking: an empirical investigation. Rev Econ Stat 72:211–218

Antonioli B, Filippini M (2001) The use of a variable cost function in the regulation of the Italian water industry. Utilities Policy 10:181–187

Anwandter L, Ozuna T (2002) Can public sector reforms improve the efficiency of public water utilities? Environ Dev Econ 7(4):687–700

Arocena P, Oliveros D (2012) The efficiency of state-owned and privatized firms: does ownership make a difference? Int J Prod Econ 140:457–465

Ashton JK (2000) Total factor productivity growth and technical change in the water and sewerage industry. Serv Ind J 20(4):121–130

Bakker K (2003) From public to private to…public? Re-regulating and mutualising private water supply in England and Wales. Geoforum 34(3):359–374

References

Banker R, Charnes A, Cooper W (1984) Some models for estimating technical and scale inefficiencies in data envelopment analysis. Manage Sci 30(9):1078–1092

Berg SV, Marques RC (2011) Quantitative studies of water and sanitation utilities: a literature survey. Water Policy 13(5):591–606

Bhattacharyya A, Harris T, Narayanan R, Raffiee K (1995) Specification and estimation of the effect of ownership on the economic efficiency of the water utilities. Reg Sci Urban Econ 25:759–784

Boardman A, Vining A (1989) Ownership and performance in competitive environments: a comparison of the performance of private, mixed and state-owned enterprises. J Law Econ 32(1):1–33

Boubakri N, Cosset J, Saffar W (2008) Political connections of newly privatized firms. J Corp Financ 14:654–673

Boycko M, Shleifer A, Vishny RW (1996) A theory of privatization. Econ J 106:309–319

Bozec R, Dia M (2007) Board structure and firm technical efficiency: evidence from Canadian State-Owned Entesrprises. Eur J Oper Res 177(3):1734–1750

Brockett PL, Golany B (1996) Using rank statistics for determining programmatic efficiency differences in data envelopment analysis. Manage Sci 42:466–472

Byrnes P, Grosskopf S, Hayes K (1986) Efficiency and ownership: further evidence. Rev Econ Stat 668:337–341

Carrozza C (2011) Italian water services reform from 1994 to 2008: decisional rounds and local modes of governance. Water Policy 13(6):751–768

Carvalho P, Marques RC (2011) The influence of the operational environment on the efficiency of water utilities. J Environ Manage 92:2698–2707

Carvalho P, Marques RC (2014) Computing economies of vertical integration, economies of scope and economies of scale using partial frontier nonparametric methods. Eur J Oper Res 234(1):292–307

Caves WC, Christensen LR, Swanson JA (1981) Productivity growth, scale economies, and capacity utilization in U.S. railroads, 1955–74. Am Econ Rev 71:994–1002

Charnes A, Cooper W, Rhodes E (1978) Measuring the efficiency of decision making units. Eur J Oper Res 2(6):429–444

Chirkos TN, Sears AM (1994) Technical efficiency and the competitive behaviour of hospitals. Socio Econ Plann Sci 28:219–227

Coelli T (1996) A guide to DEAP Version 2.1: a data envelopment analysis (computer) program. In: CEPA working paper 96/08. Departments of Econometrics, University of New England, Armidale, Australia

Coelli T (1998) A multi-stage methodology for the solution of orientated DEA models. Oper Res Lett 23:143–149

Correia T, Marques R (2011) Performance of Portuguese water utilities: how do ownership, size, diversification and vertical integration relate to efficiency? Water Policy 13(3):343–361

Cruz N, Marques R (2012) Mixed companies and local governance: no man can serve two masters. Public Adm 90(3):737–758

Cruz N, Marques R, Romano G, Guerrini A (2012) Measuring the efficiency of water utilities: a cross-national comparison between Portugal and Italy. Water Policy 14(5):841–853

Cubbin J, Tzanidakis G (1998) Regression versus data envelopment analysis for efficiency measurement: an application to the England and Wales regulated water industry. Utilities Policy 7:75–85

De Witte K, Marques RC (2010) Designing performance incentives, an International benchmark study in the water sector. CEJOR 18:189–220

De Witte K, Marques RC (2011) Big and beautiful? On non-parametrically measuring scale economies in non-convex technologies. J Prod Anal 35:213–226

Dewenter KL, Malatesta PH (2001) State-owned and privately owned firms: an empirical analysis of profitability, leverage, and labor intensity. Am Econ Rev 91(1):320–334

Dietsch M, Weill L (1999) Les performances des banques de dépots francaises: une evaluation par la méthod DEA. In: Badillo PY, Paradi JC (eds) La Méthod DEA. Hermes Science Publications, Paris

Dinc S, Gupta N (2011) The decision to privatize: finance and politics. J Financ LXVI(1):241–269

Eckel C, Vining A (1985) Elements of a theory of mixed enterprise. Scott J Polit Econ 32(1):82–94

Estache A, Kouassi E (2002) Sector organization, governance and the inefficiency of African water utilities. In: Policy research working paper no. 2890. The World Bank, Washington, DC, USA

Fabbri P, Fraquelli G (2000) Costs and structure of technology in the Italian water industry. Empirica 27:65–82

Faccio M (2006) Politically connected firms. Am Econ Rev 96(1):369–386

Faccio M (2010) Differences between politically connected and non-connected firms: a cross country analysis. Financ Manage 39(3):905–927

Fan JPH, Wong TJ, Zhang T (2007) Politically-connected CEOs, corporate governance and post-IPO performance of China's newly partially privatized firms. J Financ Econ 84:330–357

Filippini M, Hrovatin N, Zori J (2008) Cost efficiency of slovenian water distribution utilities: an application of stochastic Frontier methods. J Prod Anal 29(2):169–182

Ford J, Warford J (1969) Cost functions for the water industry. J Ind Econ 18(1):53–63

Fraquelli G, Giandrone R (2003) Reforming the wastewater treatment sector in Italy: implications of plant size, structure and scale economics. Water Resour Res 39(10):1293

Fraquelli G, Piacenza M, Vannoni D (2004) Scope and scale economies in multi- utilities: evidence from gas, water and electricity combinations. Appl Econ 36(18):2045–2057

García S, Moreaux M, Reynaud A (2007) Measuring economies of vertical integration in network industries: an application to the water sector. Int J Ind Organ 25:791–820

García-Sánchez IM (2006) Efficiency measurement in Spanish local government: the case of municipal water services. Rev Policy Res 23(2):355–371

Goldman E, Rocholl J, So J (2009) Do politically connected boards affect firm value? Rev Financ Stud 22(6):2331–2360

González-Gómez F, García-Rubio MA (2008) Efficiency in the management of urban water services. What have we learned after four decades of research? Hacienda Pública Española 185(2):39–67

Guerrini A, Romano G, Campedelli B (2011) Factors affecting the performance of water utility companies. Int J Public Sector Manag 24(6):543–566

Guerrini A, Romano G, Campedelli B (2013) Economies of scale, scope, and density in the Italian water sector: A two-stage data envelopment analysis approach. Water Resour Manage 27(13):4559–4578

Guerrini A, Romano G, Martini M (2014) Determinants of efficiency in Danish wastewater utilities. In: Working paper accepted at ECOstp IWA conference, Verona

Gupta N (2005) Partial privatisation and firm performance. J Financ 15:987–1015

Hall D (2001) Water privatisation and quality of service. Public services international research usnit. University of Greenwich, London. Available from: http://www.psiru.org. Accessed 31 July 2012

Hall D, Lobina E (2012) Financing water and sanitation: public realities. Public services international research unit. University of Greenwich, London http://www.psiru.org. Accessed 31 July 2012

Hassanein AAG, Khalifa RA (2007) Financial and operational performance indicators applied to public and private water and wastewater utilities. Eng Constr Arch Manage 14(5):479–492

Hoff A (2007) Second stage DEA: comparison of approaches for modelling the DEA score. Eur J Oper Res 181:425–435

Hunt L, Lynk E (1995) Privatization and efficiency in the UK water industry: an empirical analysis. Oxf Rev Econ Stat 57(3):371–388

Idelovitch E, Klas R (1997) Private sector participation in water supply and sanitation in Latin America. World Bank, Washington

Kim H, Clark R (1988) Economies of scale and scope in water supply. Reg Sci Urban Econ 27(2):163–183

References

Kirkpatrick C, Parker D, Zhang Y (2006) An empirical analysis of state and private-sector provision of water services in Africa. World Bank Econ Rev 20(1):143–163

Knapp M (1978) Economies of scale in sewerage purification and disposal. J Ind Econ 27(2):163–183

Li W, Xu LC (2004) The impact of privatisation and competition in the telecommunications sector around the world. J Law Econ 47:395–430

Lobina E, Hall D (2007) Experience with private sector participation in Grenoble, France, and lessons on strengthening public water operations. Utilities Policy 15:93–109

Lynk E (1993) Privatisation, joint production and the comparative efficiencies of private and public ownership: the UK water industry case. Fiscal Stud 14:98–116

Marques RC, De Witte K (2011) Is big better? On scale and scope economies in the Portuguese water sector. Econ Model 28(3):1009–1016

Martins R, Fortunato A, Coelho F (2006) Cost structure of the Portuguese water industry: a cubic cost function application. Universidade de Coimbra, GEMF

Massarutto A, Paccagnan V, Linares E (2008) Private management and public finance in the Italian water industry: a marriage of convenience? Water Resour Res 44:1–17. doi:10.1029/2007WR006443

Megginson, W.L., Nash, R.C., & Van Randenbourgh, M. (1994). The financial and operating performance of newly privatized firms: an international empirical analysis. J Financ. XLIX(2), 403-452

Ménard C, Saussier S (2000) Contractual choice and performance. The case of water supply in France. Revue d'économie industrielle 92:385–404

Menozzi A, Gutiérrez Urtiaga M, Vannoni D (2011) Board composition, political connections, and performance in state-owned enterprises. Ind Corp Change 21(3):671–698

Mizutani F, Urakami T (2001) Identifying network density and scale economies for Japanese water supply organizations. Pap Reg Sci 80(2):211–230

Nauges C, Van den Berg C (2008) Economies of density, scale and scope in the water supply and sewerage sector: a study of four developing and transition economies. J Regul Econ 34(2):144–163

Niessen A, Ruenzi S (2010) Political connectedness and firm performance. Evidence from Germany. Ger Econ Rev 11(4):441–464

Peda P, Grossi G, Liik M (2013) Do ownership and size affect the performance of water utilities? Evidence from Estonian municipalities. J Manage Gov 17(2):237–259

Picazo-Tadeo AJ, Gonzàlez-Gòmez F, Sàez-Fernàndez FJ (2009a) Accounting for operating environments in measuring water utilities' managerial efficiency. Serv Ind J 29:761–773

Picazo-Tadeo AJ, Sàez-Fernàndez FJ, Gonzàlez-Gòmez F (2009b) The role of environmental factors in water utilities' technical efficiency. Empirical evidence from Spanish companies. Appl Econ 41:615–628

Ray SC (1991) Resource-use efficiency in public schools: a study of Connecticut data. Manage Sci 37:1620–1628

Renzetti S, Dupont D (2009) Measuring the technical efficiency of municipal water suppliers: the role of environmental factors. Land Econ 85(4):627–636

Romano G, Guerrini A (2011) Measuring and comparing the efficiency of water utility companies: a data envelopment analysis approach. Utilities Policy 19(3):202–209

Romano G, Guerrini A, Vernizzi S (2013) Ownership, investment policies and funding choices of Italian water utilities: an empirical analysis. Water Resour Manage 27(9):3409–3419

Rossi D, Young E, Epp D (1979) The cost impact of joint treatment of domestic and poultry processing wastewaters. Land Econ 55(4):444–459

Saal D, Parker D (2000) The impact of privatization and regulation on the water and sewerage industry in England and Wales: a translog cost function model. Manag Decis Econ 21(6):253–268

Saal D, Parker D, Weyman-Jones T (2007) Determining the contribution of technical efficiency and scale change to productivity growth in the privatized English and Welsh water and sewerage industry: 1985–2000. J Prod Anal 28:127–139

Saal D, Arocena P, Maziotis A, Triebs T (2013) Scale and scope economies and the efficient vertical and horizontal configuration of the water industry: a survey of the literature. Rev Netw Econ 12(1):93–129

Sauer J (2005) Economies of scale and firm size optimum in rural water supply. Water Resour Res 41:1–13

Seroa da Motta R, Moreira A (2006) Efficiency and regulation in the sanitation sector in Brazil. Utilities Policy 14(3):185–195

Sexton TR, Sleeper S, Taggart RE Jr (1994) Improving pupil transportation in North Carolina. Interfaces 24:87–103

Shaoul J (1997) A critical financial analysis of the performance of privatize industries: the case of the water industry in England and Wales. Crit Perspect Account 8:479–505

Shih JS, Harrington W, Pizer WA, Gillingham K (2006) Economies of scale in community water systems. J Am Water Works Assoc 98(9):100–108

Shleifer A, Vishny RW (1994) Politicians and Firms. Quart J Econ 109:995–1025

Shleifer A (1998) State versus private ownership. J Econ Perspect 12:133–150

Simar L, Wilson PW (2004) Performance of the bootstrap for DEA estimators and iterating the principle. In: Cooper WW, Seiford LM, Zhu J (eds) Handbook on data envelopment analysis. Kluwer Academic Publishers, Boston, pp 265–298 (Chapter 10)

Simar L, Wilson PW (2007) Estimation and inference in two-sage semi-parametric models of production processes. J Econometrics 136:31–64

Sørensen RJ (2007) Does dispersed ownership impair efficiency? The case of refuse collection in Norway. Public Adm 85(4):1045–1058

Stanton KR (2002) Trends in relationship lending and factors affecting relationship lending efficiency. J Bank Financ 26:127–152

Stone, Webster Consultants for OFWAT (2004) Investigation into evidence for economies of scale in the water and sewerage industry in England and Wales. Final report

Torres M, Morrison-Paul CJ (2006) Driving forces for consolidation or fragmentation of the US water utility industry: a cost function approach with endogenous output. J Urban Econ 59:104–120

Tupper H, Resende M (2004) Efficiency and regulatory issues in the Brazilian water and sewerage sector: an empirical study. Utilities Policy 12:29–40

Tynan N, Kingdom B (2005) Optimal size for utilities? Public policy for the private sector, World Bank Note 283

Vinnari EM, Hukka JJ (2007) Great expectations, tiny benefits e Decision-making in the privatization of Tallinn water. Utilities Policy 15:78–85

Yamout G, Jamali D (2007) A critical assessment of a proposed public private partnership (PPP) for the management of water services in Lebanon. Water Resour Manage 21(3):6

Chapter 4
Investments Policies and Funding Choices

4.1 Investments Realization and Infrastructures Needs

Decrying inadequate sewage treatment, high water losses, service interruptions, and nonpotable water, Istat (2012), FederUtility (2013), and AEEG (2013) have denounced the water service infrastructures state of emergency that has a single cause and a common solution: investments.

For many years, the EU has defined targets for the quality, efficiency, and profitability of water management and has imposed standards for public health and environmental protection that the Italian infrastructure system has been unable to meet. Currently, 7 % of the Italian population is not served by a sewerage system (see Table 4.1). Even more problematic, the coverage rate for wastewater treatment is under 80 % of inhabitants, though this indicator has slowly increased from 66.4 % in 1999 to 78.5 % in 2008 (Co.N.Vi.RI 2011; ISTAT 2009).

This represents a permanent environmental regulation infringement. Italy has been convicted several times for failure to comply with European legislation (such as Directive 91/271/EC) in more than 1,100 urban agglomerations. Its lack of wastewater treatment exposes Italy to onerous penalties and jeopardizes the fulfillment of Europe's 2015 water quality objectives (2000/60/EC).

Many critical issues remain unresolved even in water supply. Continuity in water distribution is not guaranteed in several areas of the country. Although the robustness of the supply system has consistently improved over the last few decades, reducing the number of at-risk inhabitants, 9.3 % of households complained of distribution interruptions in 2011, peaking at 30 % in the south. Another critical water network issue is water losses (see Table 4.2). The National Institute of Statistics reported that the difference between the water pumped into the network and the quantity actually sold has widened from 28.5 % in 1999 to 32.1 % in 2008, with losses reaching around 40 % in the south (ISTAT 2009). These data are partially consistent with those provided by Co.N.Vi.R.I. (2011), though the former estimated average water losses at 36:55 % in the south, 39 % in the center, and 34 % in the north.

Section 4.1 was written by Giorgia Ronco, Sect. 4.2 was written by Andrea Guastamacchia, while the others sections were written by Andrea Guerrini.

Table 4.1 Coverage of sewerage and wastewater treatment

	Coverage (%)	Deficit (%)	Coverage (%)	Deficit (%)
North	94.8	5.2	84.9	15.1
Center	92.6	7.4	78.9	21.1
South	90.9	9.1	68.6	31.4
Italy	93.1	6.9	78.5	21.5

Table 4.2 Water losses, ISTAT 2009

	Water injected into distribution networks	Water supplied from the distribution networks	Percentage of total water supplied by water introduced into the distribution networks (%)	Water losses (%)
North	3.695.788	2.727.048	73.8	26.2
Center	1.661.711	1.126.674	67.8	32.2
South	2.786.014	1.679.660	60.3	39.7
Italy	8.143.513	5.533.382	67.9	32.1

To complete the picture outlined above, we must highlight that, more than 10 years after the water quality legislation (Italian Decree 31/2001 enforcing Directive 98/83/CE) came into force, some regions are still not complying with the stringent drinking water requirements prescribed.

A significant investment in new mains and treatment plants is required to overcome this critical situation. Particularly, heavy investments should be made to renew older assets reaching the end of their useful lives. The average age of Italy's water and wastewater network is 30, with a peak of 50 in the southern regions and in Lazio (Gilardoni and Marangoni 2004). It should also be noted that about half of current investment spending goes to extraordinary maintenance, making funds for network extensions and new plants scarce: this is the real cause of Italy's high water loss and poor efficiency.

As mentioned in Chap. 2, the Italian water sector was fully reformed in 1994 by the Galli Law, moving from a subsidized municipal approach to an optimized territory-level approach for infrastructure planning, service regulation, and management. The reform also used the FCR rule as a vehicle for finding the required financial resources to boost the realization of investments. However, the results achieved so far are unsatisfactory.

As politicians' main aim was keeping tariffs low, local regulators often underestimated their real investments needs and kept costs low. However, the level of investment actually realized is lower than the estimations, as is demonstrated by comparing the gap between the planned and realized per capita investments. The ATO plan provides for a total requirement of over 65 billion euros over 30 years, corresponding to an annual average of €2.2 billion, or about 37 €/inhabitant/year (BlueBook 2011). These figures are significantly undersized when compared with those of other Western countries where the capital spending for water infrastructure

4.1 Investments Realization and Infrastructures Needs

reaches an incidence on GDP between 0.35 and 1.2 % per year. Although the Galli Law increased investments from 17 €/inhabitant/year in the 1990s to more than 30 €/inhabitant/year over the last 4 years (IEFE 2012), Italy is ranked among the last countries in Europe for investments in the water sector: investment spending in other European countries is between 80 and 120 €/inhabitant/year (OECD 2006; see Fig. 4.1).

More recently, research conducted by FederUtility (2013) on a sample of 120 water companies covering over 80 % of the country shows that Italy should invest €4–5 billion per year in water services to comply with international standards, corresponding to about 80 €/inhabitant/year (see Fig. 4.2). The study shows that Italy's total 2011 investment was about 1.6 billion euros, 1.3 billion financed with

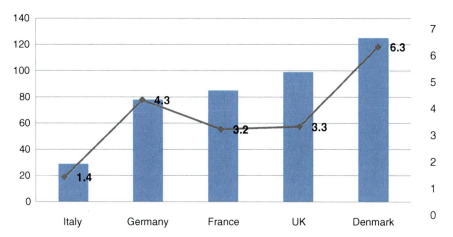

Fig. 4.1 Comparing per capita investments

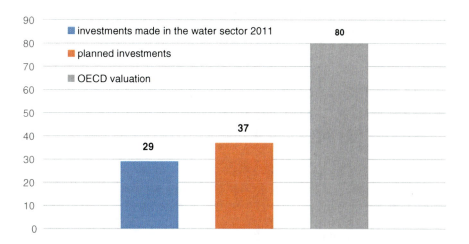

Fig. 4.2 Investment realization and infrastructure needs

tariff revenues and 0.3 billion covered by public funds. The national average per capita investment was 29 €/inhabitant, with an average public share of 6 €/inhabitant. Furthermore, the analysis reveals a striking regional imbalance: the south is far below the national average, at 16 €/inhabitant, the center is at 36 €/inhabitant and the north at 33 €/inhabitant. However, public funds reach a peak of 50 % of realized investments in the south (at 8 €/inhabitant), whereas the center and north are each at 25 %, equivalent to 4 €/inhabitant.

The FederUtility survey also shows that, on average, water utilities' willingness to make investments depends on firm size, technical and technological expertise, and their degree of profitability. Larger operators are above the national average in terms of per capita investment and are responsible for 56 % of total realized investments. The medium and small firm clusters include operators that have realized many investments and have little capital expenditure capacity, often caused by the poor industrialization of water management in small municipalities.

Furthermore, small operators were often subjected to the "CIPE tariff regime" that gave local regulators the power to impose investment scheduling: in such cases, the average investments per capita was 11 €/inhabitant in 2011, whereas, in areas were the Galli Law was fully enforced through reliable investment plans, suitable contractual agreements, and an effective tariff method, firms realized an average per capita investment of 39 €/inhabitant.

These data show that the reform begun in 1994 with the Galli Law improved investments per capita: where the reform was fully enforced, economies of scale and vertical integration were obtained, increasing the amount of investments realized. Nevertheless, further improvements are essential to overcome the challenging conditions affecting the Italian SII. First, firms' ability to obtain funds must be strengthened through an appropriate tariff method that ensures profitability and solvency. This issue will be discussed in the next paragraph describing the case of Acque Veronesi s.c.a r.l. Second, a policy of using mergers and acquisitions to achieve economies of scale and scope must be adopted, as will be discussed from paragraph 4.3 onwards.

4.2 Factors Limiting the Investment Realizations in the Italian Water Sector: The Experience of Acque Veronesi s.c.a r.l

The gap between the planned and realized investments for the reduction of network losses and the extension of sewerage and wastewater treatment was analyzed in the previous sections. Below, we provide some key figures to provide a clearer picture of the Italian water industry.

According to the OECD (2013), Italy needs an investment in water services of 65 billion euros over the next 30 years—at least €2.2 billion per year. The AEEG and FederUtility (an association of more than 400 Italian utility firms) announced

4.2 Factors Limiting the Investment Realizations in the Italian Water Sector... 41

that more than 15 billion in investments will be necessary over the next 3 years, then 20 billion over 5 years, then almost €4 billion per year.

The AEEG (2013) studied a sample of 46 companies and showed that the ratio of realized to planned investments was only 55 % at the end of 2011. This data indicate a delay of nearly €5.8 billion in investment to be realized, which, theoretically, should be added to the 20 billion needed over the next 5 years. In addition, comparing the average annual investments per inhabitant in Italy to those of other industrialized countries shows that Italy spent €26 on average per inhabitant compared to the 40 that were planned—almost half of what was spent on average in other OECD countries (OECD 2013). There are two main causes for this gap:

- The uncertainties in the regulatory framework make it difficult for water utilities to collect bank funds;
- The AATO plans promise poor profitability and require the realization of investments deemed too expensive by companies.

The AEEG (2013) divides the financing strategies to be realized by water utilities into three main categories: (a) cost reduction, (b) increased revenues (to be obtained using the so-called "3Ts"—taxes, tariffs, and transfers, and b) using repayable financing, such as bank loans and public funds.

As regards point (b), the European regulatory framework (2000/60/EC) states the following: "Member States shall take into account the principle of recovery of costs of water services, including environmental and resource costs." In addition, COM (2000) 477 stipulates that the costs fully covered by water rates include the following: (a) the financial costs of water services, including charges related to the provision of the services. They embrace all operating costs, maintenance and capital costs; (b) the environmental costs, i.e. the costs of the environmental damages caused by water resources use (such as injuries of aquatic ecosystems quality or the salinization and degradation of productive land); (c) the cost of resources, or the costs of lost opportunities for other uses as a result of the intensive exploitation of water beyond the capability of restoration and natural replacement (e.g., due to an excessive extraction of groundwater).

The European regulation was adopted in Italy through art. 154 of Law 152/2006, which states, "Tariff is based on the quality of water and of the service provided, realized investments, operating costs, costs for environmental restoration, as well as a portion of the AATO's operating costs. This method is consistent with the Full Cost Recovery (FCR) rule." This rule, recognized by two judgments of the Italian Supreme Court (335/2008 and 26/2011), requires the SII to be based on policies fostering sound economics and efficiency and requires tariffs to assure coverage of all costs up to a given threshold to avoid opportunistic firm behavior.

In 2012, the AEEG was given the authority to define the tariff method. Its first acts were the transitional tariff method (MTT) and the later new method (MTI), in force during the first regulatory period from 2012 to 2015. Before the MTT, the tariff included operational expenses (OPEX) and capital expenditures (CAPEX). Borrowing costs, income taxes, and provisions for bad credit could be charged up to 7 % of the net fixed assets. This regulatory system, called the "normalized tariff method" (MTN), only partially complies with the FCR rule, since operating costs could not go beyond the 7 %

rate; consequently, many Italian water utilities operated under poor financial conditions and had to collect funds through the abovementioned alternatives (a) and (c).

This also happened to Acque Veronesi s.c.a r.l. (AV), created in 2006 in compliance with Law 152/2006. A totally public company, AV obtained an "in-house" license to provide water services to more than seventy municipalities in the province of Verona, in the north of Italy. The firm had little equity, approximately 1 million Euros, did not own any assets such as mains or treatment plants, and was entirely controlled by municipalities and public operators. The company began operating in 2007, pursuing the realization of a plan with more than 700 million in investments over 25 years (until 2031). The lack of assets with which to calculate the MTN's 7 % rate generated a low tariff and poor cash flows, which could not properly support the realization of the plan.

The firm thus had no alternative but project financing. Unfortunately, between 2008 and 2011, the financial market was hit by a global crisis, while the water service industry was in uncertainty pending the outcome of the referendum of June 2011. Despite these difficulties, AV took out 70 million, allowing it to renegotiate its debt and make further investments. From 2007 to 2013, AV invested approximately 100 million in water mains, sewerage, and wastewater treatment plants.

The two main features of project financing are (a) the capacity of cash flows to repay the debt and its costs, keeping a positive net income and (b) the quality of the guarantees extended to safeguard the lenders at the end of the concession. The firm's loans were primarily obtained through the modification of the license agreement held with the AATO and the lending banks, which are now contemplating water utility restoration to obtain the residual value of the realized investments not yet fully covered through tariffs; this right is certainly a significant guarantee for lenders. Second, AATO Veronese's application of the FCR rule instead of the MTN allowed AV to generate higher cash flows, facilitating debt reimbursement. As mentioned, the strict application of the MTN requires that financial expenses, taxes, and losses from bad credit be covered by tariffs up to 7 % of the average net investment in the current year. This percentage, established by law in 1996, did not consider the rising interest rates, the 1997 introduction of the IRAP tax, or the increasing losses on bad credit fuelled by the economic crisis. Consequently, the 7 % rate didn't guarantee a tariff adequate for ensuring AV's solvency. An exemption from this provision was granted by AATO Veronese, giving the firm consistent aid.

Nevertheless, a question remained. The 2011 loan reimbursement plan was shorter than the useful life of the net assets during which investments were covered though tariffs. As the loan term was imposed by harsh market conditions, during a full credit-crunch period, its time horizon was 12 years, while the average reimbursement term of the investment was 16 years. Thus, all cash flows must be used to pay back the project financing, leaving no possibility for new investment. Therefore, AV was forced to cut new investment in 2013, bringing it to its lowest level in 7 years.

As AV tackled these water management problems, the AEEG issued the MTT covering 2012 and 2013, which recognized as specific components of the tariff the OPEX, CAPEX, other operating costs, borrowing costs, income taxes, and IRAP, while applying the FCR rule. Despite this improvement, the transitory method exhibited four key limits: (1) interest and income tax continued to be charged on tariffs as a percentage of investments; (2) the CAPEX was estimated

4.2 Factors Limiting the Investment Realizations in the Italian Water Sector... 43

by considering the cumulative investments realized up to 2 years prior; (3) the so-called "gradual mechanism" for efficiency improvement was problematic; and (4) the AEEG extended the useful life of investments, diluting the cash flows generated from depreciation over additional years. These provisions reduced net working capital, since companies must pay for 2 years of investments in advance before obtaining the related cash flows from tariffs, impeding solvency and profitability.

The gradual mechanism was based on the assumption that any variance between actual and planned costs was automatically explained by efficiency variations. In fact, variances could have many causes not directly related to efficiency, such as trends in production prices, the provision of services to new areas, or changes in legislation or regulatory requirements. The gradual mechanism as implemented in Acque Veronesi damaged their profitability, since unfavorable cost variances cannot be charged through the tariff. Thus, this atypical mechanism curtailed the CAPEX and OPEX covered by the tariff.

This problem was only partially compensated by the provision introducing a new tariff component, the "new investments fund" (FoNI), intended to create a financial advance for investment realization. The FoNI, conceived as an additional tariff component to cover investments, has been used to cover the unfavorable CAPEX and OPEX variances not fully covered by the tariff.

In December 2013, AEEG enacted the MTI to set the tariffs for 2014 and 2015 and confirm those for 2012 and 2013. As the MTI is still being debated, any decision could be reviewed in the ensuing months.

The regulatory framework is similar to that of the MTT, keeping unchanged the tariff recognition of the CAPEX and OPEX but adding the cost for bad credit and the environmental restoration of the water resources in order to give full effect to the FCR. One positive innovation is the introduction of two alternatives for generating investments: the first is the partial revival of the FoNI mechanism, though it retains its problems; the second is the use of financial depreciation over a shorter asset lifespan, which is more aligned with the time horizons over which finances should be reimbursed.

However, some critical issues with the MTI remain: (1) the methodology provided to quantify the financial and tax charges, the formulas for which have remained largely unchanged since the MTT, with an evident cost underestimation; (2) the underestimation of the amount of assets considered for tariff estimation; and (3) the lack of permission to direct adequate cash flows towards debt reduction. The impact of the MTI on AV will be evaluated over the next months, but the situation is likely to remain gloomy for AV and others Italian water utilities.

In addition to the new tariff method, the Italian SII also needs further reforms to improve firm solvency and boost investments:

- a modification of the "3 T" principle by introducing revolving funds and public warranties;
- a funding campaign for the water sector conducted by the Italian government and EU along with the Cassa Depositi e Prestiti and the European Investment Bank to finance long-term investments and/or offer warranties for public utilities;
- clear rules for quantifying the reimbursement of bank loans in case of the early termination or expiry of the concession.

4.3 Investment Policies and Funding Choices in the Water Sector: The Need of an Empirical Survey

The delivery of a service such as water requires costly infrastructures that are essential to the welfare of citizens and the economic development of countries (Brenneman and Kerf 2002; Briceño-Garmendia et al. 2004). Achieving an adequate level of investment is a key issue not only for developing countries but also for countries in which water scarcity, seasonality, and water leakages are significant problems. The water industry is capital intensive, with a ratio of fixed assets to annual tariff revenue of 10:1, compared to 3:1 for telecommunications, and 4:1 for electricity (Hassanein and Khalefa 2007).

A number of scholars (e.g., Idelovitch and Klas 1997; Yamout and Jamali 2007) and international organizations (e.g., OECD and the World Bank) support water industry privatization, arguing that the funding of water and wastewater utilities exceeds the capabilities of the public sector and that privatization represents a promising solution to the water supply problem. Bitrán and Valenzuela (2003) found that private utilities in Chile were better able to meet the investment needs of a highly capital-intensive sector such as the water industry: through analyses of real annual capital expenditure, the authors showed that private firms invested more than state-owned companies, partially due to their bigger size. Conversely, Hall and Lobina (2006) reported that, despite the considerable recent emphasis on privatization, private sector participation has had a negative impact on the level of investment in both developing regions (i.e., sub-Saharan Africa, South Asia, and East Asia) and developed countries. In South Asia, no investments to extend water distribution systems have been made by private water firms; moreover, in the areas analyzed, though new household connections have been made through the investments of private utilities, the number is far below expectations. The same finding was reported by Vinnari and Hukka (2007) for Estonia, where the privatization of Tallinn's water utility increased its debt exposure and tariffs.

An empirical study (Hassanein and Khalifa 2007) analyzed the debt-to-equity ratio of water utilities operating in various countries (the USA, the UK, Egypt, and other developing countries) and found that, in developing countries and Egypt, water utilities had a higher debt-to-equity ratio than in the USA, highlighting the dependence of the former areas on debt as a method of finance. Moreover, the authors found that private US water utilities had the highest debt-to-equity ratio, which was also higher than that of public US utilities, while UK utilities (all private) had a relatively balanced ratio.

In 1989, the Thatcher government privatized the regional companies in England and Wales that managed the water services, while, in Scotland and Northern Ireland, water remained controlled and operated by public authorities. When the companies were privatized, they had almost no debts, since the government had written them off. They were expected to be financed through shareholder investments, supplemented by debt through bond issuing or bank loans. Instead, the firms suffered sharp and steady debt increases and an actual reduction in shareholder equity (Hall and Lobina 2007). According to the Office of Water Services (Ofwat), the regulator of the water industry in England and Wales, there was a sharp increase in investment of about £55 billion in the 15 years after privatization, for an average of £3.7 billion per year, compared to an average of £2 billion per year during the 1980s (Ofwat 2005). This was partly due to the forced achievement of higher drinking water and wastewater treatments standards, established by the European directives requiring an averaging investment of £0.6 billion per year from 1990 (Hall and Lobina 2007). However, as highlighted by Hall and Lobina (2007), between 1985 and 1989, investment increased at a rate of 8 % per year compared to 3 % from 1989 to 2004. Moreover, the privatized companies failed to reduce leakages, which reached the levels of Eastern European and Asian cities (Hall and Lobina, 2007). Accordingly, Shaoul (1997) analyzed the privatized water industries in England and Wales and found very inadequate spending on renewal (about 1.5 % by value of infrastructure assets was spent on maintaining infrastructure, as opposed to the required 6–12 %).

In their study of the French water industry, Ménard and Saussier (2000) found that the decision to outsource water services depended on the existence of financial constraints. They also found that the larger the population, the smaller the per capita investment and the greater the profitability for operators. In such cases, private operators have an incentive to bid, since they can reasonably expect to amortize their investments within the duration of the contract. Moreover, the authors found that different water qualities or water origins requiring diverse investments (such as raw or underground water, where sparsely populated areas required much more investment) encouraged direct management by public bodies to avoid the opportunistic behavior of private operators and the resulting negative effects on water quality and health.

The above considerations suggest that the low levels of investment should also be related to the water utilities' low levels of capitalization and the associated difficulties in accessing bank loans. Moreover, some authors (Massarutto et al. 2008) have argued that the cost of capital is a relevant variable in defining investment policies.

The Italian water industry is typically associated with a low level of investment (Fabbri and Fraquelli 2000; AEEG 2013). Indeed, Italian water industry investments have decreased progressively since the 1980s (Ermano 2012). As a result of inadequate investments, leakages accounted for around 36 % of the water fed into the water grid (OECD 2013), to a maximum of 43 %, on average, in the south (Cittadinanza Attiva, 2013). Eurostat data (2009) suggest that Italy's total freshwater abstraction by public water supply is the highest in Europe.

Co.n.vi.r.i. data (2011) showed that, from 1999 to 2009, around 5.6 billion euros of investment were realized by Italy's water utilities; on a yearly basis, these investments were only a part of the planned investments (56 % in 2007, 60 % in 2008, and 61.6 % in 2009). Furthermore, while 69 % of planned investments financed by water tariffs were realized in both 2008 and 2009, only 39 % and 43.5 % (respectively) of planned investments funded by grants were carried out. Using ownership information, Co.n.vi.r.i. (2011) reported that mixed-ownership companies seem to have completed more planned investments than have public and private water utilities (over 80 %, compared to 50 and 28.8 % respectively). Moreover, mixed-ownership firms have completed more investments funded by grants than public and private companies have.

Guerrini et al. (2011) applied the financial ratio model to analyze two financial indicators (i.e., variation of tangible and intangible assets and tangible and intangible assets to population served). They found that public companies invested more than mixed-ownership firms while also applying lower tariffs. Moreover, Romano et al. (2013) show that public utilities have a higher propensity to invest in water mains, wastewater networks, and sewerage plants than private and mixed firms do. Guerrini and Romano (2013) show that the availability of bank loans and the cost of debt were crucial to water utilities' investment decisions. Moreover, Guerrini et al. (2011) found that private Italian utilities used financial leverage more intensively than public firms did. Interestingly, Massarutto et al. (2008) argue that regulation and competition, rather than ownership, are the main drivers of water utilities' efficiency and the main factors in market risk and return.

This literature review highlights the need to further investigate the factors that affect the investment and financing decisions of water utilities, in order to give policy and decision makers relevant information with which to define their strategies and make their choices. This empirical study attempts to answer the following research question: do operational and environmental variables affect the investments and financial structure of water firms? Drawing from a pioneering study (Romano et al. 2013), using financial items integrated with other technical measures such as population served and main lengths, we conducted an empirical analysis to determine the effects of some key variables, broadly studied in previous studies, on investment levels and capital costs.

4.4 Data Collection and Method of Analysis

A total of 98 firms were observed in an efficiency analysis over 5 years (2008–2012) in this empirical study on investments and funding choices. The selection process is carefully described in the method section of Chap. 3.

The data used in the analysis are on the populations these utilities served, their main lengths, net tangible assets, total assets, shareholder equity, cost of debt, and net working capital. Data on the lengths of the mains and the number of inhabitants served were generally available from websites or corporate financial statements; we also solicited information directly from company technical staff. The Bureau Van Dick AIDA database provided us with data pertaining to investments, equity, and debts. Four further indictors were then estimated. The first two are "net tangible assets per capita" and "net tangible assets per kilometers of mains length," which provided significant measures of the relevance of the investments made by water utilities. Then, the "degree of financial dependence" was calculated as the ratio of debts to total assets, providing information about the firm's exposure to banks and financiers. Finally, a "financial risk" index was determined by multiplying the degree of financial dependence for the cost of debts. A high value for this index indicates a high and expensive financial exposure, and consequently a riskier situation, for the firm.

4.4 Data Collection and Method of Analysis

The sum of the invested capital for the 98 utilities over 5 years is about 25 billion euros. The annual mean net tangible attests value is 52.7 million, while the total assets—including intangible, financial, and current assets—is 147.5 million. Table 4.3 presents a composite picture of the panel analyzed, showing the wide differences for each variable observed. Some water utilities realized more than 900 million euros of investments over 5 years and others just a few thousands of net tangible assets.

The financial structure is quite dependent on bank loans (75 %), with a 5 % interest. One of the most widely used key indicators of companies' financial health is net working capital (Altman, 1968; Hill et al. 2010). In this study, if net working capital was greater than 0, solvency was assumed to be strong; if net working capital was lower than 0, solvency was assumed to be weak. Italian water utilities show a weak solvency, with an average net working capital lower than 1 million euros.

Following the method chosen for the efficiency analysis carried out in Chap. 3, we divided the data set on investments and funding choices into groups according to four criteria.

First, we categorized companies into large, medium and small clusters according to the abovementioned EU production value parameter (firms with more than 50 million euros in sales are "large," those with between 50 and 10 million euros are "medium-sized," and those with less than 10 million are "small"). Then, firms were grouped according to their localization on the Italian peninsula (i.e., north, center, and south) Two clusters were created based on ownership structure: public firms, generally controlled by one or more municipalities, and the rest—water utilities with either mixed or private ownership. Finally, based on the ratio of the population served to kilometers of mains, we identified three approximately equally sized groups based on customer density: high density (HD; $>= 153$ nhab./km), medium density (MD; 153 inhab./km $<>$ 86 inhab./km), low density (LD; $<= 86$ inhab./km).

Table 4.4 provides an overview of the clusters, along with their descriptive statistics for investments and financial structure. The clusters differ substantially. Some firms are 24 times smaller than others when measured by net tangible investments. Mixed and private firms are, on average, larger than public utilities when net tangible investments are considered, but the latter exhibits higher value per capita, perhaps implying that public control maximizes investments. Firms operating in the center of Italy display the highest total and per capita investments. Finally, firms operating in sparsely populated areas are smaller than the MD and HD clusters but display the highest investments per capita. This interesting result implies that less densely areas require much larger investments.

The third measure, "net tangible assets per km of mains," follows the same distribution of "net tangible assets per capita" when size, localization, and ownership are considered. In low-density areas, high investments per capita coexist with low investments per km of mains; thus, improving mains in these areas is cheaper, but the burden of investments and CAPEX per capita is higher than it is in more populated regions.

Table 4.3 Brief descriptive statistics

	Net tangible assets	Total assets	Shareholders' equity	Cost of debt (%)	Net working capital	Degree of financial dependence	Net tangible assets per capita	Net tangible assets per km of mains	Financial risk
Mean	52,738,084	147,531,753	34,213,187	5	−1,057,912	75	196	23,218	0.031
Max	948,477,022	2,378,161,000	708,425,695	19	257,028,006	112	1,361	130,234	0.189
Min	–	306,737	−388,091	–	−503,120,599	13	–	–	–
St dev	100,837,202	319,113,022	80,443,126	3	51,762,462	20	252	27,716	0.029

4.4 Data Collection and Method of Analysis

Table 4.4 Descriptive statistics per cluster

Average value				
Size	Net tangible assets	Net tangible assets per capita	Net tangible assets per km of mains	Financial risk
Large	148,199,519	169	27,296	0.035
Medium	48,173,348	266	31,650	0.032
Small	6,882,517	126	10,937	0.027
Localization				
North	36,389,042	201	22,344	0.028
Center	113,050,464	235	25,465	0.036
South	46,970,573	154	23,631	0.033
Ownership				
Public	52,424,482	245	27,520	0.029
Mixed and private	53,291,931	108	15,621	0.034
Cluster density				
High density	73,660,195	125	31,515	0.030
Medium density	50,750,977	197	22,974	0.031
Low density	33,682,475	265	15,114	0.032

Table 4.5 Evolution of investments and finance

Year	Net tangible assets	Net tangible assets per capita	Net tangible assets per km of mains	Financial risk
2008	46,723,077	191	22,054	0.034
2009	49,619,130	191	22,548	0.031
2010	52,415,126	196	23,184	0.025
2011	53,325,834	194	23,399	0.032
2012	61,307,925	205	24,847	0.033

Regarding financial risk, clusters that maximize total investments are in the worst conditions when size, localization, and ownership are observed. In terms of density, the LD cluster displays the highest financial risk.

Keeping in mind that, since 2008, EU nations have been affected by a deep and widespread economic and financial crisis, particularly in Mediterranean countries (i.e., Portugal, Italy, Greece, and Spain), it is interesting to observe how water utilities' investments and finances have changed. The data in Table 4.5 show that investments progressively increased during the period observed, in both absolute and per capita terms. After an initial reduction, financial risk increased, reaching the 2008 level, through the extensive use of debt and, above all, the rising interest rates required by Italian banks.

The evidence shown with the descriptive statistics have been tested with a robust statistical model. First, we compare the means, medians, and variances for two measures: "net tangible assets per capita" and "financial risk." If the differences are statistically significant, the variable used to group the firms is a relevant determinant of investments and financial performance.

We thus applied the median, t-tests and Mann-Whitney test to reveal the differences between the two clusters created on the basis of ownership (i.e., public and mixed-private utilities). The median and Bartlett's test indicated differences across groupings based on size (i.e., large, medium, or small), density (HD, MD, or LD), and geographical localization (north, center, or south). Next, we used a regression model to verify the findings and explore the causal relationships.

The model related each measure to four independent variables:

- Production value (PV), a continuous variable measuring firm size to detect the presence of scale economies.
- Customer density (CD), indicating the presence of economies of density in the Italian water industry, measured by the ratio of population served to kilometers of main length.
- Localization (LOC), a dummy variable reflecting the geographical area where the water utilities operate (i.e., north, center, or south).
- Ownership (OWN), a dummy variable reflecting the firm's ownership (i.e., public or mixed-private).

The models are reported through:

$$\text{NTAPC} = \beta_0 + \beta_1 \text{PV} + \beta_2 \text{CD} + \beta_3 \text{LOC} + \beta_4 \text{OWN} + \varepsilon$$
$$\text{FR} = \beta_0 + \beta_1 \text{PV} + \beta_2 \text{CD} + \beta_3 \text{LOC} + \beta_4 \text{OWN} + \varepsilon$$

where NTAPC = "net tangible assets per capita"; FR = "financial risk."

Since we worked with observations of multiple phenomena obtained over a five-year period for the same water utilities, we used a panel data regression: this model takes account of the correlations among observation for each utility during the years analyzed.

4.5 Results and Discussion

Table 4.6 summarizes the results of the parametrical and non-parametrical tests. The empirical tests show large differences among clusters when "net tangible assets per capita" and "financial risk" are considered. Public firms realize the highest investments per capita while displaying the lowest financial risk. This evidence confirms the results of prior research on Italy (Romano et al. 2013). Public firms, even medium firms, with a production value between 10 and 50 million realize the highest investments, while large utilities exhibit difficulties in dealing with banks. The results according to geographical localization confirm the evidence of FederUtility (2013), described in the first section of this chapter: there is more investment per capita in the central and northern regions, followed by the islands and southern areas. Despite having the lowest investments, the southern utilities have higher

4.5 Results and Discussion

Table 4.6 Parametric and nonparametric tests

	Net tangible assets per capita	Financial risk
Ownership		
Public	245	0.029
Mixed and private	108	0.034
T test	0.000[a]	0.055[c]
Median test	0.000[a]	0.321
Mann-Whitney	0.000[a]	0.332
Size		
Large	169	0.035
Medium	266	0.032
Small	126	0.027
Bartlett's test	0.000[a]	0.032[b]
Median test	0.000[a]	0.002[a]
Localization		
North	201	0.028
Center	235	0.036
South	154	0.033
Bartlett's test	0.000[a]	0.005[a]
Median test	0.000[a]	0.075[c]
Cluster density		
High density	125	0.030
Medium density	197	0.031
Low density	265	0.032
Bartlett's test	0.000[a]	0.000[a]
Median test	0.000[a]	0.028[b]

[a] indicate 1 %
[b] indicate 10 %
[c] indicate 10 % significance levels, respectively

financial risks scores. Finally, water utilities serving low-density populations realize the highest investments and consequently are highly dependent on banks.

The evidence shown with the descriptive statistics have been tested with a robust statistical model. First, we compare the means, medians, and variances for two measures: "net tangible assets per capita" and "financial risk." If the differences are statistically significant, the variable used to group the firms is a relevant determinant of investments and financial performance.

We thus applied the median, t-tests and Mann-Whitney test to reveal the differences between the two clusters created on the basis of ownership (i.e., public and mixed-private utilities). The median and Bartlett's test indicated differences across groupings based on size (i.e., large, medium, or small), density (HD, MD, or LD), and geographical localization (north, center, or south). Next, we used a regression model to verify the findings and explore the causal relationships see Table 4.7.

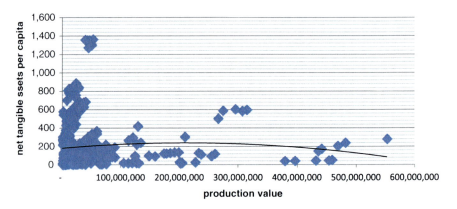

Fig. 4.3 The relationship between net tangible assets per capita and production value

Table 4.7 Evidence from regression analysis

	Net tangible assets per capita	Financial risk
Production value	0.000[b]	0.000
Density	−0.332[b]	−0.0001
Localization		
-center	92.52	0.003
-south	−33.8	0.004
Ownership		
Mixed-private	−160.95[a]	0.004

[a] indicate 1 %
[b] indicate 10 %
[c] indicate 10 % significance levels, respectively

To the best of our knowledge, ours is the first study to observe the effects of firm size and density on the amount of realized investments per capita. The former operational variable positively affects technical and global efficiency, as highlighted in Chap. 3, while also improving investment, but only up to a given threshold (Fig. 4.3).

This last result could be explained with reference to two main issues: (1) unlike large firms, small and medium enterprises are willing to maximize the quality of their services through an appropriate investment policy; and (2) large firms benefit from their size, thus achieving economies in terms of investments. The first point is not shown by the research, but the second could be partially explained by the association between large and high-density firms, as demonstrated by the chi square test reported in Table 4.7: this implies that big water utilities usually operate in high-density areas and vice versa; thus, density is one of the determinants of large firms' lower per capita investments (Table 4.8).

4.5 Results and Discussion

Table 4.8 The association between size and density

	HD	MD	LD	Total
Large	41	32	20	93
Medium	68	80	59	207
Small	49	46	78	173
Total	158	158	157	473
Pearson chi2(4) = 21.2				
Pr = 0.000				

The robust evidence showing public utilities enjoying the best performance when investments per capita are measured confirms the results of prior studies (Hall and Lobina 2006; Vinnari and Hukka 2007; Shaoul 1997), particularly those on Italy (Guerrini et al. 2011; Romano et al. 2013).

References

AEEG (2013) Fabbisogno di investimenti e individuazione degli strumenti di finanziamento per il raggiungimento degli obiettivi di qualità ambientale e della risorsa idrica—Primi orientamenti. Consultation Paper, 339
Altman EI (1968) Financial ratios, discriminant analysis and the prediction of corporate bankruptcy. J Finance 23(4):589–609
Bitrán G, Valenzuela E (2003) Water services in Chile: comparing private and public performance. Public Policy Private Sector 255. http://rru.worldbank.org/documents/publicpolicyjournal/255Bitra-031103.pdf Accessed 27 Jan 2014
Blue Book (2011) I dati sul servizio idrico integrato. Utilitatis
Brenneman A, Kerf M (2002) Infrastructure and poverty linkages: a literature review. The World Bank. http://www.ilo.org/wcmsp5/groups/public/—ed_emp/—emp_policy/—invest/documents/publication/wcms_asist_8281.pdf Accessed 9 Jan 2014
Briceno-Garmendia C, Estache A, Shafik N (2004) Infrastructure services in developing countries: access, quality, costs, and policy reform. World Bank policy research working paper 3468. World Bank, Washington
Cittadinanza Attiva (2013) Il Servizio Idrico Integrato. Indagine a cura dell'Osservatorio prezzi e tariffe di Cittadinanzattiva. www.cittadinanzattiva.it
Co.n.vi.r.i (2011) Rapporto sullo Stato dei Servizi Idrici. Comitato per la Vigilanza sull'Uso delle Risorse Idriche, Roma
Ermano, P (2012) Gli investimenti nel servizio idrico in Italia: un'analisi storica, working paper DIES 3/20. http://www.dies.uniud.it/wpdies.html?download=147 Accessed 18 Jan 2014
Fabbri P, Fraquelli G (2000) Costs and structure of technology in the Italian water industry. Empirica 27(1):65–82
FederUtility (2013) Dossier sugli investimenti
Gilardoni A, Marangoni A (2004) Il settore idrico italiano. Strategie e modelli di business. Franco Angeli, Milano
Guerrini A, Romano G (2013) The process of tariff setting in an unstable legal framework: an Italian case study. Utilities Policy 24:78–85
Guerrini A, Romano G, Campedelli B (2011) Factors affecting the performance of water utility companies. Int J Public Sector Manag 24(6):543–566
Hall D, Lobina E (2006) Pipe dreams. The failure of the private sector to invest in water services in developing countries. Public services international research unit, PSIRU Business School, University of Greenwich

Hall D, Lobina E (2007) From a private past to a public future?-The problems in England and Wales. Public Services International Research Unit, PSIRU Business School, University of Greenwich

Hassanein AAG, Khalifa RA (2007) Financial and operational performance indicators applied to public and private water and wastewater utilities. Eng Constr Archit Manage 14(5):479–492

Hill MD, Kelly GW, Highfield MJ (2010) Net operating working capital behaviour: a first look. Financ Manage 39(2):783–805

Idelovitch E, Klas R (1997) Private sector participation in water supply and sanitation in Latin America. World Bank, Washington

IEFE (2012) La riforma della regolazione dei servizi idrici in Italia. L'impatto della riforma: 1994-2011. Bocconi, Milano

ISTAT (2009) Censimento delle risorse idriche a uso civile. Roma

ISTAT (2012) Giornata mondiale dell'acqua. Roma

Massarutto A, Paccagnan V, Linares E (2008) Private management and public finance in the Italian water industry: A marriage of convenience? Water Resour Res 44:1–17

Ménard C, Saussier S (2000) Contractual choice and performance. The case of water supply in France. Revue d'économie industrielle 92:385–404

OECD (2013) OECD Environmental performance review: Italy 2013

OECD (2006) The impacts of change on the long-term future demand for water sector infrastructure. infrastructure to 2030: telecom, land transport, water and electricity

OFWAT (2005) Financial performance and expenditure of the water companies in england and wales. 2004–2005 report

Romano G, Guerrini A, Vernizzi S (2013) Ownership, investment policy and funding choices of Italian water utilities: an empirical analysis. Water Resour Manage 27(9):3409–3419

Shaoul J (1997) A critical financial analysis of the performance of privatize industries: the case of the water industry in England and Wales. Crit Perspect Account 8:479–505

Vinnari EM, Hukka JJ (2007) Great expectations, tiny benefits e decision-making in the privatization of Tallinn water. Utilities Policy 15:78–85

Yamout G, Jamali D (2007) A critical assessment of a proposed public private partnership (PPP) for the management of water services in Lebanon. Water Resour Manage 21(3):6

Chapter 5
Water Demand Management and Sustainability

5.1 Sustainable Use and Management of Water Resources: A Brief Overview

Sustainable development is a core objective of the EU, whose Water Framework Directive (2000/60/EC) is based on the idea that water management needs to take account of economic, ecological, and social issues and that its prime objective is the sustainable use and management of water resources. According to the World Commission on Environment and Development (also known as the "Brundtland Commission 1987"), sustainable activities are those where the needs of the present generation are met without compromising the needs of future generations. Specifically, sustainable water use is "the use of water that supports the ability of human society to endure and flourish into the indefinite future without undermining the integrity of the hydrological cycle or the ecological systems that depend on it" (Gleick et al. 1995).

Recent research suggests that the vast majority of the world's population (nearly 80 %) is highly exposed to water security threats and that even the richest nations are not trying to remedy the causes of this dangerous situation (Vörösmarty et al. 2010).

Water resources will soon come under further pressure through many factors, such as population growth and urbanization, economic development, and climate change (Beck and Bernauer 2011; Güneralp and Seto 2008; Serageldin 2007; United Nations 2009; Vörösmarty et al. 2000). The EU is becoming increasingly concerned about drought events and water scarcity; a growing number of EU Member States have experienced seasonal or longer term droughts and water scarcities. Policymakers must therefore balance the increasing human demand for water with the protection of ecosystem sustainability.

Sections 5.1, 5.2.4 and 5.3 were written by Giulia Romano, Sects. 5.2.1, 5.2.2 and 5.2.3 were written by Martina Martini, while Sect. 5.4 was written by Francesco Fatone.

© The Author(s) 2014
A. Guerrini and G. Romano, *Water Management in Italy*, SpringerBriefs in Water Science and Technology, DOI 10.1007/978-3-319-07818-2_5

The three main users of water are agriculture, industry, and what is referred to as the "domestic" sector, including households and services. As highlighted by Olmstead (2010), water conservation generally refers to the technical water savings that can be achieved through a particular technology or policy intervention.

Water utility conservation policies can use various instruments. The most relevant and commonly used are water pricing, incentives for the implementation of high-efficiency appliances, rationing policies, and information campaigns designed to improve public awareness of the activities that are useful in reducing water consumption. Implementing measures to reduce water demand can deliver benefits not only at the economic level but also at the environmental and social ones (Dworak et al. 2007).

Residential customers account for most of the water demand in urban areas, mainly through household appliances such as baths, showers, toilets, dishwashers, and washing machines. Domestic water demand management may help reduce water shortages and lessen the growing pressure on the environment. It may also reduce the necessity for the construction of major infrastructure, thus reducing the need for new investments and lowering costs (March and Saurì 2009). Thus, a deep knowledge of household users' consumption behavior is crucial for policy makers and utilities managers.

Since consumer requests and expectations of corporate environmental sustainability are increasing (Veleva 2010) and a number of environmental problems, including water scarcity, are caused by consumer lifestyles, raising awareness of water conservation and the daily activities that can reduce water consumption is necessary (Dworak et al. 2007).

Protecting the environment and sustainably managing natural resources such as water are among the broad activities supported by the EU Horizon 2020 framework. Moreover, at an international level, UNESCO (the United Nations Educational, Scientific, and Cultural Organization) addressed the water issue by implementing the International Hydrological Programme (IHP) in 1975 and then building upon it on three tracks, one of which is water resources assessment and management to achieve water sustainability. The current phase (IHP-VIII) is titled "Water security: Responses to local, regional, and global challenges." It will run from 2014 to 2021 and will focus on six themes, among which water education is considered the key to water security.

In some countries, for example, water utilities use websites to promote sustainable practices. For instance, besides emphasizing the importance of daily actions such as taking a shower rather than a bath, flushing as seldom as possible, turning off the taps while brushing teeth, making full use of the dishwasher, using bowls to wash dishes and vegetables, installing dual flush toilets, and reaching a full load before using the washing machine, water firms in England and Wales use their websites to promote water conservation using devices such as water butts, tap inserts, shower flow regulators, and save-a-flush products. Households are informed that they can capture the water falling onto their roofs by installing water butts, containers in which the water can be recovered. This gives families another source of water without having to tap into their home supply; rainwater is also better for plants. Tap inserts save many liters of water per day if they are inserted into washbasin taps; they mix air with the water, giving the same effect while using much less water. Furthermore, installing a shower flow regulator allows a high-quality shower while saving water and money through the control of the flow rate, making it as low as possible, ensuring that the right amount of water is delivered to the shower. The save-a-flush bags are filled with crystals that expand in the toilet cisterns; they save one liter of water per flush and can be used in old single-flush toilets. Some of these products are offered by English and Welsh water utilities to their customers for free, making households more keen to adopt them. Another device promoted by companies is the shower timer, which can be used while showering by turning the timer over and getting out when the sand runs out; it lets people know how much time they spend in the shower, making then more keen to take their showers faster.

Most empirical studies on household conservation policies focus on a particular area of the USA or Australia (e.g., Lee et al. 2011; Barrett and Wallace 2011; Fielding et al. 2013).

5.2 Policies for Sustainable Water Use: A Review of the Literature

5.2.1 Tariff Policy

Given the economics of water, price can be considered one of the main tools for managing water demand and promoting equity, efficiency, and sustainability in the water sector (Rogers et al. 2002). Various schemes are used to set water prices, which can be divided into three main models: a constant unit price, an increasing block rate (in which the average price, constant within the same block, increases by raising the block of consumption), and the decreasing block rate (in which the average price, constant within the same block, decreases by increasing the block of consumption). In addition to the variable component, linked to the volume used, total expenditures for water should include a fixed component, as a payment for the service. These two components, a tax on volumes and a fixed fee for access, are preconditions for allocative efficiency and are related to the consumer surplus, which in turn consists of two parts—the surplus resulting from the consumption and the surplus arising from the connection (Sibly 2006).

The design of a tariff system can accomplish up to four purposes (Dalhuisen and Nijkamp 2002; Howe 2005): full cost recovery, economic efficiency, equity, administrative feasibility, and tariff system efficiency. Full cost recovery is necessary to support a utility's operations and maintenance and to cover its debt costs, opportunity costs, and environmental externalities (Rogers et al. 2002). Concerning economic efficiency, a tariff should promote the reduction of consumption while providing incentives for realizing investments: tariffs thus play a key role in both demand and supply. The equity purpose implies that water prices

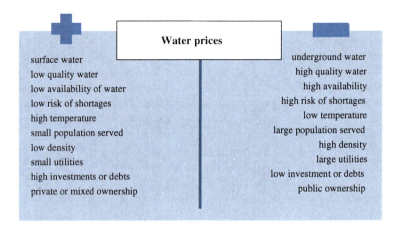

Fig. 5.1 Factors affecting water prices

must ensure equitable access to the resource and to its use and future availability. Thus, water prices can only partially be used to reduce water demand, because an increase in price manly affects low-income households, which must allocate a larger part of their budget to water expenditures or give up essential uses; lower prices are not necessary fairer, however, since they worsen the scarcity problem, harming future generations, particularly the poorest (Bithas 2008). Finally, these three purposes must be reached by utilities at the minimum cost, thus ensuring the administrative feasibility and efficiency of the tariff system.

Not only their purposes but also other relevant factors affect water prices. These variables, widely investigated in previous studies, can be divided into five main groups: those concerning water characteristics, weather conditions, population characteristics, consumption, and utilities policy (see Fig. 5.1).

Concerning water characteristics, using underground water, which carries lower treatment costs than using surface water, is associated with lower prices (Thorsten et al. 2009; Ruester and Zschille 2010). Higher water quality and availability reduce prices (Martinez-Espiñeira et al. 2009; Zetland and Gasson 2013) and the risk of water shortages because of the low costs associated with poor-quality services (Zetland and Gasson 2013).

Regarding weather, high temperatures and scarce rainfall lead to higher prices by increasing water demand and reducing water availability (Martinez-Espiñeira et al. 2009; Thorsten et al. 2009). The same relationship is found in studies that compare tariffs applied in regions with different climatic conditions (Martinez-Espiñeira et al. 2009).

Population features also affect water prices. Providing the service to a larger number of users or in more densely populated areas allows economies of scale and density, which in turn result in lower prices (Martinez-Espiñeira et al. 2009; Ruester and Zschille 2010; Zetland and Gasson 2013). The evidence for economic conditions is less consistent: higher prices are sometimes associated with lower incomes and higher poverty rates and sometimes with higher economic well-being (Martinez-Espiñeira et al. 2009; Thorsten et al. 2009). Low levels of consumption are related to high prices (Zetland and Gasson 2013).

Finally, significant relationships have been found between utilities' policies and prices. Large firms apply lower tariffs, confirming the existence of economies of scale (Thorsten et al. 2009), while high investments and high debts increase prices (Marin 2009; Thorsten et al. 2009; Ruester and Zschille 2010). Finally, private and mixed-ownership firms apply higher tariffs than public ones (Saal and Parker 2001; Bitrán and Valenzuela 2003; Kouanda and Moudassir 2007; Marin 2009; Martinez-Espiñeira et al. 2009; De Witte and Saal 2010; Ruester and Zschille 2010; Guerrini et al. 2011).

Prices and their differentiation according to uses, income levels, areas, and seasons influence consumption and thus utilities' revenues (Howe 2005). The relationship between water prices and consumption has been investigated to define the strengths and weaknesses of price as a tool for managing water demand. The pioneering study is Howe and Linaweaver (1967), which demonstrated that water demand is substantially inelastic to price; an empirical analysis carried out in the Unites States using the regression technique revealed an elasticity of -0.23. The same model was later applied by Gibbs (1978) and Foster and Beattie (1979) to evaluate the price elasticities in Miami and the United States (respectively): both studies showed coefficients slightly higher than -0.5.

Other studies examine the effects of price used jointly with other water demand management tools. Nieswiadomy (1992) carried out an analysis in 1984 on the use of price along with conservation and educational programs in the United States, finding that water demand had a low elasticity to marginal price (-0.17) and that the other programs, whose effects were limited to drier areas, were even less effective. Michelsen et al. (1999), using US panel data spanning 11 years, compared price policies to nonprice policies, including information and educational programs, modernization policies, and decrees and rules. The study reveals a price elasticity of -0.23, reductions of between 1.1 and 4 % attained by each program, and the absence of a significant relationship between price and nonprice policies. Similarly, Renwick and Archibald (1998) analyzed how prices, restrictions, and conservative technologies affected water demand in California from 1985 to 1990, revealing that a 10 % increase in price caused a 3.3 % reduction in water demand compared to the greater effectiveness of the other two tools. Higher price elasticity values are seen in the analysis of Wang et al. (1999), carried out in the United States from 1992 to 1997, on price, information campaigns, and water savings technologies used as integrated strategies by a local utility to reduce water demand: price elasticity increased from -0.508 to -0.697 through growing differences in price among blocks of consumption volume.

Coefficients of elasticity resulting from these studies show that water demand is inelastic to price, suggesting the partial effectiveness of price policies as tools for reducing water consumption. However, as highlighted by Renwick and Green (2000), inelasticity does not mean that water demand is unreactive to price, given that water use varies according to price, albeit less than proportionally and in relation to the influence of other variables.

First, price elasticity is affected by *seasons*: water demand is more elastic in summer due to the presence of discretionary uses, mainly outdoor ones (such as watering the garden), which can be given up after a cost–benefit analysis. This was demonstrated by Renwick and Green (2000) in a study carried out in California from 1989 to 1996: they found an elasticity to marginal price of -0.16, with a demand that was more reactive in summer (by about 25 %) than in winter and in large areas subject to irrigation.

A second factor in price elasticity is the *time variable*: water demand is generally more reactive to price in the long run than in the short term. This was confirmed by Nauges and Thomas (2003), who analyzed water demand in France from 1988 to 1993, using the generalized method of moments (GMM). They found that price elasticity in the short run was lower than it was over a period of 6 years (-0.26 to -0.40, respectively); the differences were explained with reference to technologies, which constrain consumption levels and cannot easily be replaced, and the habits guiding consumer behaviors. Musolesi and Nosvelli (2007) applied the same method to estimate the water demand in 102 Italian municipalities from 1998 to 2001. They observed a water demand elasticity of -0.27 in the short term, lower than that in the long run (-0.47).

Differences in price elasticity should also be related to the *tariff system*, with the resulting marginal and average prices and the consequent consumer perception. Olmestead et al. (2007), using a sample of 1.082 Canadian and U.S. families, showed higher price elasticity in increasing bock rates than in situations characterized by a constant marginal price. Nieswiadomy and Molina (1991) obtained similar results in Texas. Applying a tariff scheme based on a decreasing block rate from 1976 to 1980 and a tariff scheme based on an increasing block rate from 1981 to 1985, to 101 consumers, they found that consumers reacted more to average prices in the first period and more to marginal prices in the second, thus concluding that consumer perception affects behavior. The importance of the consumer perception also emerges in Martins and Maura e Sa (2011). Using a questionnaire administered to 386 Madeira families in 2008, they demonstrated how comprehension of the tariff scheme influenced the efficacy of price signals, since only the consumers who understood the causal relation between volume used and water bill adopted environmentally friendly behaviors. They concluded that price is an effective tool for managing water demand only if it is used jointly with information policies.

The method used in an analysis also influences the elasticity appraisal, as demonstrated by Hewitt and Hanemann (1995). They applied the discrete/continuous choice model to the Nieswiadomy and Molina (1991) database, finding a price elasticity of -1.6, higher than that estimated by previous studies using OLS. The influence of the method was also demonstrated by Pint (1999), who used the heterogeneous preferences model and the two-error model to estimate the water demand of 599 single-user households in California from 1982 to 1992. The first method revealed a price elasticity variable from -0.04 to -0.14 in summer and from -0.20 to -0.29 in winter, while the second model found a variable from -0.20 to -0.40 in summer and -0.33 to -1.24 in winter. These results not only suggest that elasticity value depends on the method used but also differ from those obtained by Renwick and Green (2000) regarding the effect of seasons on price elasticity.

Finally, elasticity is influenced by *income*. Water is a normal good since its consumption increases as income increases, though less than proportionally. Several studies have proved this positive relationship: Renwick and Archibald (1998) showed an income elasticity of water demand in California from 1985 to 1999 of 0.36, similar to that estimated in Germany in 2003 by Scleich and Hillenbrand (2009). Lower values, around 0.10, were found in Siviglia from 1991 to 1999 and in France from 1988 to 1993, respectively, by Martínez-Espiñeira and Nauges (2004) and Nauges and Thomas (2000). Finally, a middle value, 0.18, was found in Italy from 1998 to 2001 by Musolesi and Nosvelli (2007).

Our analysis does not focus on income elasticity per se but on the influence of income on the price elasticity of water demand. Previous studies offer significant insights into the equity issues related to the use of price as a tool for managing water demand. In particular, Renwick and Archibald (1998) and Renwick and Green (2000) show that low-income families react more to price increases because water expenditures have a high incidence on their income; price policies are thus more successful in communities characterized by a high proportion of low-income people. Similarly, Agthe and Billings (1987), using a sample of families in Arizona grouped by income, found that poor families use less water and are more reactive to price increases; they concluded that effective and fair water policies are based on price increases that become progressively higher in the upper income blocks, equaling the marginal utility of the poor and wealthy. Similar conclusions were drawn by Ruijs et al. (2008) in a study carried out in Brazil from 1997 to 2002. They estimate that poor families spend from 4.2 to 4.7 % of their income on water, while the wealthy spend only 0.4 to 0.5 % despite consuming more than twice as much. Thus, price increases based on consumption volume or income level are more fair but have high administrative costs (especially those based on income), and they reduce water utility revenues. For these reasons, the authors suggest combining price with no-price policies to manage water demand fairly and effectively.

The influences of these variables—season, time, tariff scheme, method of analysis, and income—and of the other factors affecting price elasticity estimates were finally investigated by three meta-analyses that sum up the findings of several studies on this topic (see Table 5.1).

The first meta-analysis, by Espey et al. (1997), based on 124 observations from 24 studies published in the United States from 1967 to 1993, reports a price

Table 5.1 Meta-analyses on factors affecting the price elasticity of water demand

Characteristics of the sample	Espey et al. (1997)	Dalhuisen et al. (2003)	Waddams and Clayton (2010)
No. of studies	24	64	148
No. of observations	124	296	1.308
Location	United States	United States and Europe	–
Years of publication	1967–1993	1963–2001	1963–2008
Average price elasticity	−0.51	−0.41	−0.38
min; max (price elasticity)	−3.33; −0.22	−7.47; 7.90	−7.47; 3.5

elasticity between −3.33 and −0.02 (an interval that shrinks between −0.75 and 0.00 in 90 % of the studies), negatively influenced by income, evapotranspiration, and rainfall and that is higher in summer than in winter, in the long run than the short run, and with the inclusion of commercial uses and price structures different from the marginal price. The types of data and econometric model used for the estimation are not relevant.

The second meta-analysis, by Dalhuisen et al. (2003), is based on 296 observations from 64 studies, published and unpublished, covering from 1963 to 2001. It finds a price elasticity distribution with a mean equal to −0.41, a minimum of −7.47, and a maximum of 7.90. The price elasticity is positively influenced by increasing block tariffs accounting for income differences, while lower elasticity values are found in countries with high per capita income; they are also lower in Europe than in the United States, in unpublished analyses, and in studies that use the marginal price.

Finally, the third meta-analysis, by Waddams and Clayton (2010), uses 1,308 observations from 148 studies published from 1963 to 2008. It finds an average price elasticity of −0.38, within a range from −7.47 to 3.5, influenced by data characteristics and the model of demand specification. The use of panel data or annual data rather than infra-annual data reduces elasticity, while the inclusion of income, rainfall, evapotranspiration, and commercial uses positively affects price elasticity. The method applied, geographical areas, seasons, publication status, and the authors' gender do not matter.

The studies and meta-analyses indicate that elasticity estimation is influenced by various factors relating to research design and sample features; as the latter affect not only the estimated elasticity but also the real effectiveness of price policies, their analysis is instrumental in designing appropriate tariff schemes in relation to population characteristics.

After identifying the conditions that limit the effectiveness of price as a tool for managing water demand and the equity issues relating to its use among people with different income levels, the last aspect to be considered is the efficiency profile. Olmstead and Stavins (2008) argue that price policies are more efficient than nonprice policies because they allow consumers to choose their volume of water in relation to their needs and willingness to pay, assuring the best allocation of the resource in the market without any additional costs; the effectiveness of nonprice policies is subordinated to management and monitoring activities, which have huge administrative and control costs.

The same conclusions were drawn by Campbell et al. (2004), who posited the possibility of applying prices generally and with cumulative effects, despite the problems of price diversification based on income and other socioeconomic variables. Finally, Rogers et al. (2002) and Barrett (2004) offered a different, partially conflicting, point of view by stating that the efficiency assessment has to consider the externalities generated by the use of the resource and the related costs: water prices will consequently increase because of the inclusion of externality and administrative costs linked to their appraisal.

5.2 Policies for Sustainable Water Use: A Review of the Literature

Table 5.2 Strengths and weaknesses of price as a water demand policy

Strengths	Weaknesses
Effectiveness: water demand reactivity (Renwick and Green 2000; Campbell et al. 2004) Higher effectiveness in the long run (Espey et al. 1997; Nauges and Thomas 2003; Musolesi and Nosvelli 2007; Martínez-Espiñeira and Nauges 2004; Waddams and Clayton 2010)	Limited effectiveness: inelasticity of water demand (148 studies and 3 meta-analysis: Espey et al. 1997; Dalhuisen et al. 2003; Waddams and Clayton 2010)
No control costs (Olmstead and Stavins 2008)	High administrative costs (Rogers et al. 2002; Barrett 2004)
Generally and widely applied (Campbell et al. 2004)	Difficult to differentiate; increasing block tariffs (Nieswiadomy and Molina 1991; Campbell et al. 2004; Olmestead et al. 2007)
	Unfair (Agthe and Billings 1987; Espey et al. 1997; Renwick and Archibald 1998; Renwick and Green 2000; Bithas 2008; Ruijs et al. 2008)

Table 5.2 sums up the strengths and weaknesses of price as a tool for managing water demand. It highlights price's limited effectiveness, especially in the short term, with water demand only partially reactive to price variations. Regarding efficiency, price policies do not have control costs and should be generally and widely applied, letting each consumer buy volumes of water according to his budget and preferences, assuring an optimal allocation of the resource in the market; however, costs arise if administrative activities are considered, especially those related to the evaluation of externalities or price differentiation based on consumption or income levels. Finally, price increases are most burdensome for low-income households, raising equity issues concerning the use of this tool to manage water demand.

5.2.2 Rationing and Restrictions

Rationing and restrictions are two nonprice tools for managing water demand. Rationing implies service disruptions during some hours and days of the week and should follow criteria that consider the households' composition. Restrictions forbid or limit usage, applicable generally to the more discretionary uses such as irrigation in certain hours of the day or washing the yard. Both are used as emergency measures to cope with drought conditions because they have immediate effects and are usually widely applied.

The effectiveness of these policies when used jointly with others has been proven by several empirical analyses that show reductions in consumption ranging from 0 to 29 %. For example, Renwick and Archibald (1998), studying South California data covering 1985 to 1990, demonstrate the effectiveness of restrictions on irrigation used jointly with another policy. A comparative analysis of various policies is offered by Renwick and Green (2000) in another study on California covering 1989 to 1996, showing that rationing caused a 19 % reduction in water demand, while restrictions caused a 29 % reduction—both deeper than the reductions induced by information campaigns (−8 %), water saving technologies (−9 %), or price increases (−1.6 %).

Restrictive policies and information campaigns could be jointly used to reduce water demand, as is demonstrated in a 2002 analysis carried out in Virginia by Halich and Stephenson (2009), which does not just consider the presence of the program but includes information on the program content, the type of restriction, the informational effort, and the enforcement measures related to monitoring, control, and penalties. The results show that implementation intensity influences water demand: restriction programs followed by an information policy proceeding from moderate to high reduced water demand from 6 to 12 %, a range that moved from 15 to 22 % when control activities were implemented. Enforcement tools increased the effectiveness of voluntary and mandatory programs, but they had high political and staff costs that made their adoption inconvenient. Similar conclusions were drawn by Olmstead and Stavins (2008), who subordinated the effectiveness of mandatory programs to the intensity of control activities, whose costs have to be added to consumer and utilities losses in order to assess their net benefit.

While the combined use of mandatory policies and information campaigns results in steeper water demand reductions, price policies, and restrictions do not have additive effects. The positive coefficient of the interaction between the two policies seen in Kenney et al. (2008) using Colorado data covering 1997 to 2005 shows that restrictions cause consumers to react less to price. Moreover, responsiveness to price differs among consumers and in relation to drought conditions: it is higher with a higher level of consumption and after a drought, suggesting that price has to be used to manage water demand with the aim of controlling higher uses in the long run, while restrictions must be implemented to deal with drought conditions in the short term.

Confining restrictions and rationing to drought situations is also justified by their social welfare loss, when assessed alone or together, to that of price. In a study carried out in Siviglia during the early 1990s, Garcia Valiñas (2006) analyzed the welfare losses of households and companies based on consumer surplus variations in relation to various prices. The results showed that, in drought periods and without restrictions, households had higher losses than companies did, while ability to pay (especially households') and thus welfare losses increased if hour restrictions were implemented. Finally, the comparison of restriction to rationing showed that water quality was more important to households while quantity was more important to industries.

A focus on restrictions on outside uses, which represent about half of the water consumed by households in most Australian cities was offered by Brennan et al. (2007): a mild restriction was compared with an absolute prohibition of irrigation, showing that the last one is less effective and it causes higher losses for consumers.

5.2 Policies for Sustainable Water Use: A Review of the Literature

Welfare losses caused by restrictions are more evident if compared with that of price, as demonstrated by Woo (1992), Roibás et al. (2007), Grafton and Ward (2008), and Mansur and Olmstead (2012). In particular, Woo (1992) analyzed service interruptions imposed in Hong Kong in 1973–1990: interruptions are inefficient, because they limited consumers choices causing welfare losses 500 times higher than those induced by price policy. Similarly, Roibás et al. (2007) analyzed reduction in water supply and price policy as tools for rationing water in Seville during the drought of 1992–1996: regardless of the behavior adopted—reduction in the volume consumed or use of special tanks to mitigate the effect of interruptions—welfare losses associated with disruption are greater than in the hypothesis of price increases. The same authors, however, highlighted the incompleteness of the model that do not consider the reduction in water losses caused by the decrease in pressure. Even Grafton and Ward (2008), using a model that includes the additional cost of tanks bought by consumer to compensate water restrictions, demonstrated the higher welfare losses caused by restrictions than prices in Sidney in 2004–2005. Authors defined rationing inefficient, because it overlooks consumer preferences and the different value of water uses, but at the same time necessary to deals with a severe drought, inducing an immediate and temporary reduction in the volume used, or to decrease water demand if consumptions are not measured. Finally, also Mansur and Olmstead (2012) demonstrated the advantages of prices on restriction policy, using water demand of 1.082 families in the United States and Canada in 1996–1998: price allows families to choose the volume of water they want in relation to their preferences and availability to pay, so using prices instead of restrictions caused a welfare gain for families equal to 29 % of their annual expenditure for water, or even higher if benefits of innovation and diffusion of technologies are considered, generally absent in "command and control" regulations.

Finally, our analysis focuses on equity issues related to the use of rationing and restrictions. Duke et al. (2002) considered the effects of 25 % reduction in water demand, alternatively induced by price, rationing, and mandatory restrictions on outside uses. The authors defined price as a noneffective and fair tool to manage water demand, respectively because the majority of consumers is not aware of the relation between volume used and total expenditure for water and because price increases mainly affect poor households, who allocate a consistent portion of their income to water expenditure. Rationing, instead, immediately reduces water demand, even if its effectiveness should be limited by consumption increases when water is available and it bears upon poorest households, that have to give up essential uses. Mandatory restrictions on external uses are instead more fair, because they shift the burden of conservation on those who live in larger areas.

Also Barrett (2004) emphasized the higher equity of regulations rather than prices, considering the example of the introduction of restriction on irrigation that mainly affects richest households, increasing equity among people. This conclusion differs from that of Roibás et al. (2007), who stressed that cuts in supply are regressive, because richest households should buy technologies for water supply and conservation. Rationing and restrictions seem to be fairer than prices, even if

Table 5.3 Strengths and weaknesses of rationing and restrictions

Strengths	Weaknesses
Effectiveness (Renwick and Archibald 1998; Renwick and Green 2000; Barrett 2004; Halich and Stephenson 2009) High effectiveness in droughts (Kenney et al. 2008)	Costs of control and penalties (Olmstead and Stavins 2008; Halich and Stephenson 2009)
Generally and widely applied (Campbell et al. 2004)	High welfare losses (Woo 1992; Garcia Valiñas 2006; Brennan et al. 2007; Roibás et al. 2007; Grafton and Ward 2008; Mansur and Olmstead 2012)
Fairer than price (Duke et al. 2002; Barrett 2004)	Unfair (Roibás et al. 2007)

their contribution to enhance equity among households depend on their design and the associated control activities (Survis and Root 2012).

Table 5.3 sums up evidences of previous studies, highlighting strengths and weaknesses of rationing and restrictions as water demand management tools, under three profiles—effectiveness, efficiency, and equity.

5.2.3 Technology Devices

Technology devices are nonprice measures to manage water demand. They include different devices, such as mechanisms for saving water in the toilet, in the shower and in the bath, meters of water consumption, efficient appliances, rainwater harvesting systems, devices for recycling gray water (e.g., the use of discharge water from showers, baths, and sinks in the toilet flushes), and systems for the reduction of water networks losses (EA 2008).

The adoption of water saving technologies can be enabled by incentives and rebate programs or through regulatory activities, which may limit their implementation to certain conditions, such as restructuring or construction of new buildings. The final choice to invest in efficient technologies is influenced by environmental variables and consumer characteristics, such as socioeconomic variables, (e.g., age, income, and ownership), attitudinal and behavioral factors (e.g., water-saving habits), and specifications of the dwelling, including year of construction, size, number of rooms, and extent of outdoor spaces (Millock and Nauges 2010; Martínez-Espiñera and García-Valiñas 2013).

In addition to these factors, the effectiveness and payback period of the technology investment play an important role. Different studies measured water savings of technology devices. In particular, a recent analysis by Tsai et al. (2011) evaluated the effects of the adoption of four different technologies in Massachusetts: results show that the installation of control taps for irrigation, sensitive to weather conditions, reduces the variability in the use of water; rainwater harvesting systems provide a significant amount of water for outside uses, but not enough to satisfy domestic purposes; finally, devices for the control of consumption and grants for toilet and washing machines with low consumption have greater effectiveness, as well as other technologies to retain moisture in the soil. A similar analysis was carried out by Muthukumaran et al. (2011): they focused on the reuse of gray water and demonstrated its high conservative potential, but with a recycling limited to toilet flush and external uses, such as watering the garden.

Meters of the volume consumed represents another conservative technology: a visual display gives information on the level of consumption, while sometimes acoustic signals are activated when predefined thresholds, possibly disaggregated by type of use, are overcome. These monitors provide dynamic feedback to influence consumer behavior by information available in real time. The use of these devices was studied by Willis et al. (2010) in the Australian Gold Coast in 2008 using a sample of 151 households and a subsample of 44 users, which were equipped by a monitor in the shower with an acoustic signal at 40 liters of consumption. The educational potential of such devices decreases the consumption of water in the showers of 27 %, which leads to estimated savings for the whole city of 3 % in water and 2.4 % in energy; finally, the authors pointed out that the adoption of display meters can be supported by the short payback period, equal to 1.65 years, and by an annual rate of return, defined on 10 years, of 23.3 %.

The effectiveness of water saving devices, high in the first years of implementation, decreases over time because of off-setting behavior: consumers indeed get used to water efficiency of technology devices, they quantify the related water savings and so increase their consumptions, within their budget. These conclusions were drawn by Lee et al. (2011, 2013), Lee and Tansel (2013). The first study, carried out in Florida, analyzed the effects on consumption of the replacement of old showers, toilets, and washing machines with new higher efficiency mechanisms: it was demonstrated that water demand significantly decreases during the first 2 years of implementation, while water savings diminish in the third and fourth year. Similar results were achieved also by Lee and Tansel (2013): using a questionnaire administered by telephone to 271 families in Florida they analyzed the effects of the replacement of toilets, showers and aerators, showing end-users satisfaction, synergic effects resulting from the implementation of several measures and changes in consumption habits, high in the first and second years of implementation, but that disappear in the third year of observation. The analysis by Lee et al. (2013) instead, based on disaggregated data, shows a delay in consumption decrease: the largest decline in demand is recorded in the third year, while decreases are lower in the fourth year of observation and they do not characterized the upper class of consumption, which records a modest increase in water demand.

The presence of off-setting behaviors was proved by other studies that consider their use with other water demand management policies. For example, Geller et al. (1983) tried to identify a successful mixed between educational, behavioral and technological strategies in Virginia at the end of Seventies: the analysis revealed significant water savings caused by technology devices supported by educational campaigns, even if savings were lesser than expected due to the persistence of off-setting behaviors that reduce the benefits of efficient devices. A later analysis by Campbell et al. (2004), conducted in Phoenix on 6 years consumptions of 19,000 households, similarly detected off-setting behaviors, that were more pronounced in individuals who have not chosen the program independently; the authors suggested the opportunity to limit or avoid off-setting behaviors through a communication personalized on the basis of ethnicity, poverty, and age. Finally, also the literature review by Inman and Jeffrey (2006) stressed the effects of these conducts, caused by the awareness of water savings resulting from the adoption of technology devices, and the consequent need for their joint implementation with other measures.

In this perspective, Timmins (2003) proposed to support water saving technologies with price policy: the analysis, based on a simulation model applied in California in 1970–1992, did not compare benefits of price and nonprice policies, but focused on their effectiveness, demonstrating that technologies alone do not ensure a sustainable water use, but they need to be supported by appropriate pricing policy. The author finally underlined that a tax on water will increase social surplus and allow to achieve water conservation objectives, which must be distinguished from redistributive goals, that often justify the decision to keep out pricing policies.

Similar conclusions were drawn by Dawadi and Ahmad (2013), which predicted possible future scenarios in relation to different water saving measures by using a system dynamic model applied to consumptions in Las Vegas in years 1989–2035: the empirical analysis highlighted that conservation technologies on outside uses (e.g., the design of garden that require few water) have an higher conservative potential than that of technologies applied to internal uses (e.g., low-consumption appliances); furthermore this potential is greatly enhanced by the simultaneous increase of 50 % in price, that will jointly lead to an estimated reduction of 30.6 % in water demand by 2035.

Future scenario were predicted also in the study of Schwarz and Ernst (2009): based on a questionnaire administered in the South of Germany, it highlighted that the diffusion of water-saving technologies, such as efficient showers and toilets and rainwater harvesting systems, do not need promotion activities, but if these activities are undertaken, the conservation potential of technologies will enhance.

Technologies allow to achieve multiple objectives and in particular water savings, energy savings and the consequent reduction in carbon dioxide emissions. This was supported by the Environment Agency (EA 2009b), which demonstrated that in water supply-use-treatment cycle about the 89 % of carbon dioxide are emitted in the use stage (Fig. 5.2). As a consequence, optimization of systems that provide hot water in house allows to achieve not only water savings, but also significant reductions in CO_2 emissions: Hackett and Gray (2009) demonstrated that water-efficient appliances, jointly with a rational use of hot water in homes, allow to reduce water consumption and CO_2 emissions respectively by 50 % and 58 %.

The analysis of Fidar et al. (2010), finally, quantified the CO_2 emissions and the energy consumption caused by water-saving appliances in dwellings, in accordance with the English Code for Sustainable Homes (CSH). The study showed that 96 % of the energy used in houses and 87 % of CO_2 emissions are linked to

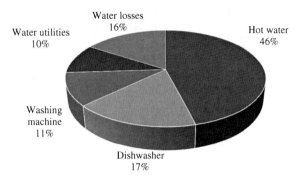

Fig. 5.2 CO_2 emissions in water supply-use-treatment cycle. *Source* EA (2009b)

water consumption, so highlighting the need to adopt an integrated approach in the choice of technologies which considers the link between water consumption, energy consumption and CO_2 emissions.

On the efficiency profile, technologies for water saving are affected by two different types of costs: R&D development costs and control costs. In relation to the first category the Environment Agency (EA 2009a) underlined the long payback period of investments that makes consumers reluctant to their adoption; similarly water utilities are not interested in promoting the implementation of such devices, since a reduction in consumption would cause a decline in revenues and profits; on supply side, the adoption of conservation technologies is made unattractive by the operators interest for short-term profits; finally, also government authority are not interesting in promoting water saving technologies, since they obtain positive judgments for their ability to keep prices low, rather than for the choice to transfer the relative costs on end users. With regard to the second costs category Olmstead and Stavins (2008) stressed the presence of administrative costs for monitoring and reinforcement activities, that assure the effectiveness of technology devices as water demand management measures.

Finally, the presence of costs associated to the implementation of technologies and their payback period on several years (Willis et al. 2013) lead to consider equity issues: poor households do not have access to technology devices, while high income household should invest in water saving appliances, supporting their costs and enjoying their benefits in the long run.

Socioeconomic variables influence the adoption of this measure, which can be encouraged by rebate programs or be made compulsory by specific legislative provisions, to which are however associated control and enforcement costs.

Table 5.4 sums up evidences from previous studies on the use of technology devices to manage water demand: in particular the effectiveness of the tool seems to be limited to short periods because of off-setting behaviors, that require their joint use with price policy or control activities. Water saving technologies also present investment costs and R&D costs that limit their attractiveness, which, on the other side, is enhanced by the possibility to achieve multiple goals, such as reductions in CO_2 emissions.

Table 5.4 Strengths and weaknesses of technology devices as water demand management policy

Strengths	Weaknesses
Effectiveness in the short run (Inman and Jeffrey 2006; Willis et al. 2010; Muthukumaran et al. 2011; Tsai et al. 2011)	Effectiveness limited by off-setting behavior (Geller et al. 1983; Campbell et al. 2004; Inman and Jeffrey 2006; Lee et al. 2011, 2013; Lee and Tansel 2013)
Effectiveness in the long run only with price policy (Timmins 2003; Dawadi and Ahmad 2013)	Costs of control (Olmstead and Stavins 2008)
Reduction of CO_2 emissions (EA 2009b; Hackett and Gray 2009; Fidar et al. 2010)	Investment costs (Willis et al. 2013) R&D costs (EA 2009a, b)

5.2.4 Information Campaigns

Information campaigns are considered important initiatives in reducing household water consumption by promoting water-saving devices or encouraging more rational water use. Delorme et al. (2003) found that consumers did not acknowledge how their own individual actions threatened community water availability and quality, tending to perceive that "other people" were more irresponsible with water than they were. Their study shows that consumers supported campaigns to educate the community on better water management practices using websites. Thus, although the authors are aware of the challenges and complexity of marketing water management, they would also like to highlight the many opportunities for promoting water management campaigns.

Information campaigns motivate households to attempt to implement more water-efficient behaviors and provide information on how to reduce usage. They could thus be considered a form of social marketing (Andreasen 1994, 1995; Stead et al. 2007a) since they seek to "influence social behaviors not to benefit the marketer but to benefit the target audience and the general society" (Kotler and Andreasen 1996, p. 389).

Water utility companies have asked consumers to renounce activities and habits they enjoy, such as baths and long showers, green gardens, clean cars (Barrett and Wallace 2011). Stead et al. (2007a, b) found that interventions that adopt marketing principles can be effective and can influence policy as well as individuals. Domene and Saurì (2006) show that consumer behavior is an important explanatory factor in household water consumption, albeit to a lesser extent than other variables (e.g., sociodemographic and economic variables, such as house type and income). Moreover, Grafton et al. (2011) show that water demand management policies that include campaigns for promoting water-saving behaviors and the use of water-saving devices would be more effective if households faced a volumetric charge for their water consumption.

Nieswiadomy (1992), Michelsen et al. (1999), Hurd (2006), Lee et al. (2011), and March et al. (2013) find that water conservation programs affect water demand. Nieswiadomy (1992) and Renwick and Green (2000) find that public

5.2 Policies for Sustainable Water Use: A Review of the Literature

education campaigns have reduced water usage. Moreover, Martínez-Espiñeira and Nauges (2004) affirm that information campaigns or promotions for low-water-using equipment should be preferred to water-pricing strategies once a given threshold of water consumption is reached.

Furthermore, as argued by Barrett (2004), although higher prices will encourage better water usage, price increases may become only a means of raising water utility revenues rather than reducing water consumption without the assistance of nonprice measures.

Romano et al. (2013) investigate the factors affecting water utility companies' decision to implement public information campaigns via corporate websites aimed at promoting sustainable water use and reducing household water consumption. They provide some interesting insights into the type of companies most sensitive to water sustainability issues: larger firms located in the center of Italy, in drought regions, and in the driest areas seem to be more eager to promote the reduction of household water consumption. Moreover, companies operating only in the water business, are publicly owned, and apply lower tariffs embody the type of institutions that make greater use of web information campaigns to reduce consumption. Moreover, Romano et al. (2013), studying both Italian and Portuguese water utilities, confirm the results of the previous Italian study; they also show that Portuguese utilities seem to be more eager than are Italian companies to promote the reduction of household water consumption through web information campaigns.

5.3 Promoting Conservation Practices of Water Use Through Web Sites: An Empirical Analysis on Italian Water Utilities

5.3.1 Data Collection and Method Adopted

Due to information campaigns' current and future importance in promoting sustainable water use, it is quite surprising to find that the literature has only marginally focused on whether water utility companies encourage the reduction of household water consumption through public information campaigns on water saving. Furthermore, reference should be made to specific factors that affect water utility companies' decision to promote these campaigns, placing due emphasis on the central role that water utility companies could play in promoting and encouraging best practices.

We consider it important to determine (Romano et al. 2013) whether Italian water utility companies' ownership influences water conservation policies, given the inherent contradiction between a water company's interest in increasing sales and water conservation (Barrett 2004). Moreover, it is important to investigate whether environmental factors or other policies encouraged by normative reforms, such as the integration of water services to exploit economies of scope or the

Table 5.5 The distribution of firms among clusters

North	Center	South	Mono-utility	Multi-utility	Large	Medium	Small	Public	Mixed-Private
195	34	75	202	102	40	82	182	162	142

merging of water utility companies to exploit economies of scale, have impacted firms' decisions to promote web saving campaigns.

To analyze Italian water utilities' willingness to promote water conservation through their web sites, we examined all the water utilities that were operating in the water industry in Italy (304 firms) at the end of 2013 according to the AEEG database. Using it along with the AIDA database, corporate websites, annual reports, and newspapers, we collect the following information on Italian water utilities:

- Ownership structure;
- Number of employees at the end of 2012;
- Diversification strategies (i.e., mono- or multi-utilities);
- Existence of a corporate web site;
- Presence on the corporate website of information on how to reduce water consumption such as taking a shower rather than a bath, flushing as seldom as possible, turning off the taps while brushing teeth, making full use of the dishwasher, using bowls to wash dishes and vegetables, installing dual flush toilets, and waiting to have a full load before using the washing machine;
- Presence on the corporate website of information about water quality (i.e., the chemical, physical, and biological characteristics of the water provided by the water utility to its customers).

The firms selected are grouped by localization (i.e., north, center, and south of Italy), scope of operations (i.e., mono or multiutility), size (i.e., large, medium, or small), and ownership structure (public or mixed-private). Table 5.5 shows that most utilities are located in the north. This demonstrates (1) the faster adoption of Galli Law provisions by northern regions, and (2) the higher fragmentation of water services in these areas (i.e., 27 % of firms are located in Lombardia), as confirmed by the predominance of small firms. By contrast, the central regions are characterized by large and medium firms under public–private control as the result of a determined regional policy of corporatization and firm concentration. Many southern firms are small mono-utilities, often controlled by private owners.

5.3.2 Results and Discussion

Table 5.6 reports that, among the selected firms, 67 (22 %) have no corporate website and thus have chosen not to use a website to provide information to their customers or, more generally, to their stakeholders. It is worth noting that 79.6 %

Table 5.6 The use of website to provide information campaign

Information on water savings		
No	163	54 %
Yes	74	24 %
No website	67	22 %
Information on water quality		
No	73	24 %
Yes	164	54 %
No website	67	22 %
Total	304	

Table 5.7 The use of websites among different clusters

	North	Center	South	Mono-utility	Multi-utility	Large	Medium	Small	Public	Mixed-Priv.
No. of firms	195	34	75	202	102	40	82	182	162	142
Information on water savings										
No	124	14	25	97	66	19	54	90	96	67
Yes	37	17	20	50	24	21	28	25	53	21
No website	34	3	30	55	12	0	0	67	13	54
Percentage (%)										
No	64 %	41 %	33 %	48 %	65 %	48 %	66 %	49 %	59 %	47 %
Yes	19 %	50 %	27 %	25 %	24 %	53 %	34 %	14 %	33 %	15 %
No website	17 %	9 %	40 %	27 %	12 %	0 %	0 %	37 %	8 %	38 %
Information on water quality										
No	50	9	14	44	29	8	16	49	35	38
Yes	111	22	31	103	61	32	66	66	114	50
No website	34	3	30	55	12	0	0	67	13	54
Percentage (%)										
No	26 %	26 %	19 %	22 %	28 %	20 %	20 %	27 %	22 %	27 %
Yes	57 %	65 %	41 %	51 %	60 %	80 %	80 %	36 %	70 %	35 %
No website	17 %	9 %	40 %	27 %	12 %	0 %	0 %	37 %	8 %	38 %

of Italy's population 11–74 years old (38.4 million people) have access to the Internet, and two-thirds of Italian families own a computer (Audiweb 2013).

Moreover, only 24 % of the 304 firms use their website as a platform to promote the sustainable consumption of water (31 % of firms with website). More attention is paid to water quality, as 164 out of the 237 firms with their own website (69 %) provide information on the physical and chemical properties of the water they deliver.

Notwithstanding the relevance of this issue at the international and normative levels (e.g., EU Directives, UNESCO), this evidence highlights that using websites, particularly to give stakeholders relevant information about sustainable consumption, is not currently important for many water utilities.

However, it is worth examining the results for each of the four clusters to better explain the evidence obtained Table 5.7.

Regarding geographical localization, water utilities operating in the central regions use websites more frequently than the others to promote water savings. The southern cluster seems to accord slight importance to disclosure and communication with customers and stakeholders through the web, as 40 % of the firms do not have a website, while the northern water utilities often do not use their website to provide information about sustainable water use.

Regarding water quality, the differences among clusters are reduced: with the exception of the south, most water utilities provide information through the Web. This similarity is also found when mono and multiutilities are observed: no relevant differences appear concerning information on water savings or on water quality. Size could be one determinant of disclosure, since all large and medium firms have a website and most provide information. Ownership structure could also be a factor in the willingness to promote water sustainability and disclosure on water quality: 33 % of public firms publish a set of standards for the reduction of water use, against the 15 % of mixed-private firms, a difference that increases if information on quality is considered (70 vs. 35 %).

Thus, these results suggest that mixed-private and small firms display the least disclosure, while large water utilities and those operating in central regions show the most intensive use of websites. To confirm these results, we report the coefficients obtained with the multinomial logit regression applied for the dependent variables observed.

The first model studies factors affecting the willingness to promote water-saving practices among citizens. Localization, size, and ownership exert relevant effects. Water utilities operating in the central regions are more likely to publish information campaigns on their website, as are large and public firms. Southern firms perform better than northern ones. Only the scope of operation is an insignificant variable.

These results are partially consistent with those in Romano et al. (2013), with the exception of those or mono- and multiutilities. Firms localized in areas characterized by water scarcity, such as the south of Italy (which also includes the two main Italian islands of Sicily and Sardinia), are encouraged to promote campaigns for water savings. The south of Italy is currently affected by drought, and its municipalities often apply water rationing policies. Fully public companies pay greater attention to this issue than do mixed or totally private firms, perhaps because a public shareholder's goal is maximizing the benefits of the community through the preservation of water resources, avoiding any waste and excess consumption. By contrast, private shareholders are more oriented toward profit and thus are less interested in decreasing water consumption, since this implies a contextual decrease in revenues and net income. Finally, the presence of some scale incentives for the implementation of sustainable water use campaigns was detected; the evidence relating to the clusters defined on the basis of employees shows that these incentives exist. Larger companies that provide water services to many citizens have a greater willingness to invest resources in water conservation campaigns, since their potential recipients are very numerous. Moreover, due to the higher number of people to reach and inform, the best tool is probably a website, since it is the most efficient and the least expensive. By contrast, the same

5.3 Promoting Conservation Practices of Water Use Through Web Sites...

Table 5.8 Evidence from regression models

Multinomial Logit		Information on water savings	Information on water quality
No	Base outcome		
Yes			
	Localization		
	–Center	1.73***	−0.003
	–South	0.98**	0.12
	Multi-utility	−0.45	−0.39
	No. employees	0.001**	0.003***
	Mixed-Private	−1.26***	−1.00***
No website			
	Localization		
	–Center	0.92	0.46
	–South	1.36**	1.24**
	Multi-utility	0.23	0.095
	No. employees	−0.08***	−0.086***
	Mixed-Private	1.30***	0.95*

***,** and * indicate 1, 5 and 10 % significance levels, respectively

campaign is not cost-effective for a company serving only a few thousand customers and that owns, on average, fewer resources that can be invested in effective websites/campaigns to encourage the reduction of household water consumption and the promotion of sustainable water use.

The apparent attitude to creating a website (see the "no web site" column of Table 5.8) shows that southern, small, and private firms are likely to operate without any Web portal on which to provide information about their activities. This shows a lack of familiarity with media and corporate communication.

Thus, water quality is a central corporate communication issue, regardless of localization and scope of operations. However, as in this case, large and public utilities outperform the others. Even though water quality is a crucial issue for users, private partners do not care to provide information on it, while small firms are likely to incur high costs and face a lack of resources when they want to update their website.

5.4 Wastewater Technologies to Reduce Environmental Impacts

Wastewater treatment (WWT) is an important link in the water cycle (Fig. 5.3).

According to the Italy's National Statistics Institute (ISTAT 2008) the current national demand for municipal WWT is higher than 81 million persons equivalent. 75.2 % of this national demand could be met by 18,000 constructed wastewater

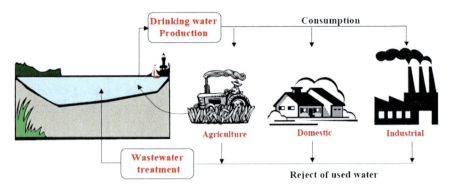

Fig. 5.3 Role of wastewater treatment in the water cycle

treatment plants (WWTPs). However, the operating WWTPs are less than the constructed and meet only 59 % of the national demand.

Further to this framework, in order to comply with the European objectives (Directive 271/91/EC), approximately 30 billion euros (46 % of all the investments required for the integrated water services) would be required over the next 30 years for WWT (source: Federutility 2012). Such a high public investment should be addressed toward sustainability and best available techniques

This approach could lead Italy to speed up the innovation in water innovation and stimulate uptake of water innovations by market and society, according to the aims of the European Innovation Partnership for Water.

In summary, to date conceiving wastewater treatment means tackling energetic, environmental and economic challenges. Apart from generating high-quality water and biosolids, new-conceived WWTPs must incorporate issues as resource recovery, energy, odors, greenhouse gases, emerging contaminants, economical efficiency, and social acceptance. Thus, the conception of sustainable WWTPs has to be based on a holistic approach, in which a plant-wide (i.e., including all the inputs and outputs), multi-disciplinary (i.e., with technical, environmental, social, and economic considerations), and flexible (i.e., depending on the geographical and socioeconomic situation) perspective is included (Fig. 5.4) (www.water2020.eu).

First of all the domestic wastewater should now being looked at more as a resource than as a waste, a resource for water, for energy, and for the plant fertilizing nutrients, nitrogen (N) and phosphorus (P).

As far as wastewater-energy nexus is concerned, WWT currently accounts for about 3 % of the electrical energy load in developed countries. In Italy conventional WWTs are energy intensive and consume about 40–50 kWh per person equivalent per year with operation expenses of about 1 billion per year only for electrical energy consumptions (Campanelli et al. 2013). On the other hand, the potential energy available in raw wastewater entering a municipal WWTP significantly exceeds the electricity requirements of the treatment process. Energy captured in organics entering the plant can be related to the organic load of the influent flow.

5.4 Wastewater Technologies to Reduce Environmental Impacts

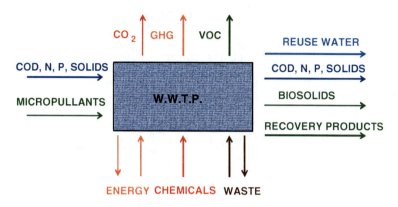

Fig. 5.4 Schematic of the new-conceived WWTPs and its interactions with the environment

Water reuse is already widely practiced where water is in limited supply, but this often increases the energy needed for treatment because of increased water quality requirements for reuse.

Reducing treatment energy requirements can help offset this need, particularly through more efficient capturing of the biofuel potential in wastewater itself. Reducing net energy requirements for wastewater treatment is a complementary, not an alternative goal to water reuse. The same can be said with respect to nutrient recovery. Additionally, climate change concerns associated with fossil fuel consumption, as well as increasing energy costs, necessitate that greater efforts be made toward better efficiency and more sustainable use of wastewater's energy potential. While more efficient water and nutrient recovery from wastewater are important goals in themselves (McCarty et al. 2011).

If more of the energy potential in wastewater were captured for use and even less were used for wastewater treatment, then wastewater treatment might become a net energy producer rather than a consumer. Energy-saving and energy-conservation measures (US-EPA 2012) together with novel bioprocesses (Metcalf and Eddy 2014) should be considered in retrofitting existing and designing future WWTPs (Fig. 5.5).

As far as the WWT-carbon nexus is concerned, major effects on climate change come from direct emissions of greenhouse gases (GHG), such as nitrous oxides (N_2O), which has a Global Warming Potential (GWP) 300 times that of CO_2 (for a 100-year timescale). N_2O emissions in municipal WWTPs are estimated in 0.22 TgN/yr (Kampschreur et al. 2009), that is about 3.2 % of the total anthropogenic emissions of nitrogen oxides. Taking into account the carbon footprint of WWT, the reduction of N_2O emissions by novel treatment processes should be fairly incentivized in the coming years.

Many chemicals and microbial agents, that were neither traditionally considered contaminants nor used for design and operation of conventional WWTPs, are now present in the environment on a global scale (Schwarzenbach et al. 2006).

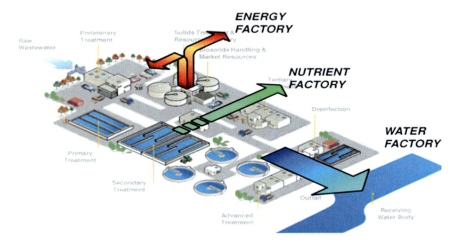

Fig. 5.5 WWTP as a factory of renewable resources

The European Water policy and legislation consider these emerging issues within the following Directives: Water Framework Directive (WFD) 2000/60/EC; Environmental Quality Standards (EQS) Directive 2008/105/EC; Directive 2009/90/EC on technical specifications for chemical analysis and monitoring of water status (QA/QC); New Directive on priority substances 2013/39/EU. On the other hand, an open debate is ongoing about the actual effectiveness of "end of pipe" measures such as the upgrading of WWTPs with comparison to the "measures at source". Many large research and monitoring projects have been carried out in Europe over the last 20 years (i.e., EU-FP6 project POSEIDON; EU-FP7 project NEPTUNE) to analyze the situation. However, only Switzerland has recently taken "end-of-pipe" actions in WWTPs. In particular, after a long term monitoring Switzerland has decreed to invest 1 billion euro to upgrade 100 (out of 750) WWTPs by application of tertiary and quaternary treatments to about 50 % of the municipal wastewater. In particular, the Swiss strategy is taking actions in: (1) WWTPs with treatment potential higher than 100,000 PE; (2) WWTPs with treatment potential higher than 30,000 PE and influencing drinking water resources; (3) WWTP with treatment potential higher than 10,000 PE at watercourses with small dilution. Such a national action could mark a turning point for a European and Italian strategy to face the chemical emerging pollution of water bodies.

As far as nutrients management is concerned, wastewater facilities are increasingly being asked or required to implement treatment process improvements in order to meet stricter discharge limits with respect to nitrogen and phosphorus; the latter has cost implications. Balancing between the nutrients removal efficiency and the cost of treatment is critical in order to implement a solution that will meet the required limits at an acceptable cost. On the other hand, major efforts should be addressed to the recovery of phosphorus which is a limited and scarce

5.4 Wastewater Technologies to Reduce Environmental Impacts

resource (Gilbert 2009). Toward these directions, novel emerging processes and technologies for nutrients removal and recovery have been developed up to pilot or demonstration scale (US-EPA 2013).

In light of such a complex scenario, Environmental Decision Support Systems (EDSS) should be used for planning and design the municipal WWT. These systems will prioritize the possible technical alternatives taking into account even the environmental and economic sustainability.

References

Aghte DE, Billings BR (1987) Equity, price elasticity, and household income under increasing block rates for water. Am J Econ Sociol 46(3):273–286

Ajzen I (1991) The theory of planned behavior. Organ Behav Hum Decis Process 50:179–211

Andreasen A (1994) Social marketing: its definition and domain. J Public Policy Mark 13(1):108–114

Andreasen A (1995) Marketing social change: changing behavior to promote health, social development, and the environment. Jossey-Bass, San Francisco, CA

Audiweb (2013) Ricerca di Base sulla diffusione dell'online in Italia e i dati di audience del mese di dicembre 2012. Available at http://www.audiweb.it/cms/view.php?id=4&cms_pk=277

Barrett G (2004) Water conservation: the role of price and regulation in residential water consumption. Econ Pap 23(3):271–285

Barrett G, Wallace M (2011) An institutional economics perspective: the impact of water provider privatisation on water conservation in England and Australia. Water Resour Manag 25:1325–1340

Beck L, Bernauer T (2011) How will combined changes in water demand and climate affect water availability in the Zambezi river basin. Global Environ Change 21:1061–1072

Bithas K (2008) The sustainable residential water use: sustainability, efficiency and social equity. The European experience. Ecol Econ 68:221–229

Bitrán GA, Valenzuela EP (2003) Water services in Chile: comparing private and public performance. World Bank, Washington

Brennan D, Tapsuwan S, Ingram G (2007) The welfare costs of urban outdoor water restrictions. Aust J Agric Res Econ 51:243–261

Campanelli M, Foladori P, Vaccari M (2013) Consumi elettrici ed efficienza energetica nel trattamento delle acque reflue. Maggioli Editore, Rimini

Campbell HE, Johnson RM, Larson EH (2004) Prices, devices, people, or rules: the relative effectiveness of policy instruments in water conservation. Rev Policy Res 21(5):637–662

Dalhuisen JM, Nijkamp P (2002) Critical factors for achieving multiple goals with water tariff systems: Combining limited data sources and expert testimony. Water Resour Res 38(7):1–11

Dalhuisen JM, Florax RJ, de Groot HL, Nijkamp P (2003) Price and income elasticities of residential water demand: a meta-analysis. Land Econ 79(2):292–308

Dawadi S, Ahmad S (2013) Evaluating the impact of demand-side management on water resources under changing climatic conditions and increasing population. J Environ Manage 114:261–275

Delorme DE, Hagen SC, Stout IJ (2003) Consumers' perspectives on water issues: directions for educational campaigns. J Environ Educ 34(2):28–35

De Witte K, Saal D (2010) Is a little sunshine all we need? On the impact of sunshine regulation on profits, productivity and prices in the Dutch drinking water sector. J Regul Econ 37(3):219–242

Domene E, Saurì D (2006) Urbanisation and water consumption: influencing factors in the metropolitan region of Barcelona. Urban Stud 43(9):1605–1623

Duke JM, Ehemann RW, Mackenzie J (2002) The distributional effects of water quantity management strategies: a spatial analysis. Review Reg Studies 32(1):19–35

Dworak T, Berglund M, Strosser P, Roussard J, Kossida M, Berbel JE, Kolberg S (2007) Final report EU Water saving potential (Part 1 –Report). Ecologic—Institute for International and European Environmental Policy

EA (2008) Greenhouse gas emissions of water supply and demand management options. Environment Agency, Bristol

EA (2009a) Evidence. A low carbon water industry in 2050. Environment Agency, Bristol

EA (2009b) Quantifying the energy and carbon effects of water saving full technical report. Environment Agency, Bristol

Espey M, Espey J, Shaw WD (1997) Price elasticity of residential demand for water: a meta-analysis. Water Resour Res 33(6):1369–1374

Fidar A, Memon FA, Butler D (2010) Environmental implications of water efficient microcomponents in residential buildings. Sci Total Environ 408:5828–5835

Fielding KS, Spinks A, Russell S, McCrea R, Stewart R, Gardner J (2013) An experimental test of voluntary strategies to promote urban water demand management. J Environ Manag 114:343–351

Foster HS, Beattie BR (1979) Urban residential demand for water in the United States. Land Econ 55:43–58

Garcia Valiñas M (2006) Analysing rationing policies: drought and its effects on urban users' welfare (Analysing rationing policies during drought). Appl Econ 38(8):955–965

Geller ES, Erickson JB, Buttram BA (1983) Attempts to promote residential water conservation with educational, behavioral and engineering strategies. Popul Environ 6(2):96–112

Gibbs K (1978) Price variable in residential water demand models. Water Resour Res 14(1):15–18

Gilbert N (2009) Environment: the disappearing nutrient. Nature 461:716–718

Gleick PH, Loh P, Gomez S, Morrison J (1995) California water 2020: a sustainable vision. Pacific Institute Report. Pacific Institute for Studies in Development, Environment, and Security. Oakland, California, USA

Grafton RQ, Ward MB (2008) Prices versus rationing: Marshallian surplus and mandatory water restrictions. Econ Record 84:57–65

Grafton RQ, Ward MB, To H, Kompas T (2011) Determinants of residential water consumption: evidence and analysis from a 10-country household survey. Water Resour Res 47:W08537

Guerrini A, Romano G, Campedelli B (2011) Factors affecting the performance of water utility companies. Int J Public Sector Manag 24(6):543–566

Güneralp B, Seto KC (2008) Environmental impacts of urban growth from an integrated dynamic perspective: a case study of Shenzhen, South China. Global Environ Change 18:720–735

Hackett MJ, Gray NF (2009) Carbon dioxide emission savings potential of household water use reduction in the UK. J Sustain Develop 2(1):36–43

Halich G, Stephenson K (2009) Effectiveness of residential water-use restrictions under varying levels of municipal effort. Land Econ 85(4):614–626

Hewitt JA, Hanemann WM (1995) A discrete/continuous choice approach to residential water demand under block rate pricing. Land Econ 71(2):173–192

Howe CW (2005) The functions, impacts and effectiveness of water pricing: evidence from the United States and Canada. Water Resour Develop 21(1):43–53

Howe CW, Linaweaver FP (1967) The impact of price on residential water demand and its relation to system design and price structure. Water Resour Res 3(1):13–32

Inman D, Jeffrey P (2006) A review of residential demand-side management tool performance and influences on implementation effectiveness. Urban Water J 3(3):127–143

ISTAT (2008) Censimento delle risorse idriche a uso civile

Kampschreur MJ, Temmink H, Kleerebezem R, Jetten M, van Loosdrecht M (2009) Nitrous oxide emission during wastewater treatment. Water Res 43:4093–4103

Kenney DS, Goemans C, Klein R, Lowrey J, Reidy K (2008) Residential water demand management: lessons from Aurora, Colorado. J Am Water Resour Assoc 44(1):192–207

Kouanda I, Moudassir M (2007) Social policies and private sector participation in water supply—The case of Burkina Faso. UNRISD project

Kotler P, Andreasen A (1996) Strategic marketing for non-profit organisations. Prentice-Hall, New Jersey

Lee M, Tansel B (2013) Water conservation quantities vs customer opinion and satisfaction with water efficient appliances in Miami, Florida. J Environ Manage 128:683–689

Lee M, Tansel B, Balbin M (2011) Influence of residential water use efficiency measures on household water demand: A four year longitudinal study. Resour Conserv Recycl 56:1–6

Lee M, Tansel B, Balbin M (2013) Urban sustainability incentives for residential water conservation: adoption of multiple high efficiency appliances. Water Resour Manage 27:2531–2540

Mansur ET, Olmstead SM (2012) The value of scarce water: Measuring the inefficiency of municipal regulations. J Urban Econ 71:332–346

March H, Saurí D (2009) What lies behind domestic water use? A review essay on the drivers of domestic water consumption. Boletín de la Asociación de Geógrafos Españoles 50:297–314

Marin P (2009) Public–private partnerships for urban water utilities: a review of experiences in developing countries. World Bank, Washington

Martínez-Espiñeira R, Nauges C (2004) Is all domestic water consumption sensitive to price control? Appl Econ 36(15):1697–1703

Martinez-Espiñeira R, García-Valiñas MA, González-Gómez F (2009) Does private management of water supply services really increase prices? An empirical analysis. Urban Stud 46(4):923–945

Martínez-Espiñera R, García-Valiñas MÁ (2013) Adopting versus adapting: adoption of water-saving technology versus water conservation habits in Spain. Int J Water Resour Dev 29(3):400–414

Martins R, Maura e Sa P (2011) Promoting sustainable residential water use: a Portuguese case study in ownership and regulation. Policy Stud 32(3):291–301

McCarty P, Bae J, Kim J (2011) Domestic wastewater treatment as a net energy producer can this be achieved? Environ Sci Technol 45:7100–7106

Metcalf & Eddy, Tchobanoglous G, Stensel HD, Tsuchihashi R, Burton F (2014) Wastewater engineering: treatment and resource recovery. McGraw-Hill, New York

Michelsen AM, McGuckin JT, Stumpf D (1999) Nonprice water conservation programs as a demand management tool. J Am Water Resour Assoc 35(3):593–602

Millock K, Nauges C (2010) Household adoption of water-efficient equipment: the role of socio economic factors, environmental attitudes and policy. Environ Resource Econ 46(4):539–565

Musolesi A, Nosvelli M (2007) Dynamics of residential water consumption in a panel of Italian municipalities. Appl Econ Lett 14:441–444

Muthukumaran S, Baskaran K, Sexton N (2011) Quantification of potable water savings by residential water conservation and reuse—a case study. Resour Conserv Recycl 55:945–952

Nauges C, Thomas A (2000) Privately operated water utilities, municipal price negotiation, and estimation of residential water demand: the case of France. Land Econ 76(1):68–85

Nauges C, Thomas A (2003) Long-run study of residential water consumption. Environ Resource Econ 26:25–43

Nieswiadomy ML (1992) Estimating urban residential water demand effects of price structure, conservation, and education. Water Resour Res 28(3):609–615

Nieswiadomy ML, Molina DJ (1991) A note on price perception in water demand. Land Econ 67(3):352–359

Olmestead SM, Michael Hanemann W, Stavins RN (2007) Water demand under alternative price structures. J Environ Econ Manage 54:181–198

Olmstead SM, Stavins RN (2008) Comparing price and non-price approaches to urban water conservation. Working Paper n. 14147, National Bureau of Economic Research

Olmstead SM (2010) The economics of managing scarce water resources. Rev Environ Econ Policy 4(2):179–198

Pint EM (1999) Household responses to increased water rates during the California drought. Land Econ 75(2):246–266

Renwick ME, Archibald SO (1998) Demand side management policies for residential water use: who bears the conservation burden? Land Econ 74(3):343–359

Renwick ME, Green RD (2000) Do residential water demand side management policies measure up? An analysis of eight California water agencies. J Environ Econ Manage 40:37–55

Rogers P, Silva RD, Bhatia R (2002) Water is an economic good: How to use prices to promote equity, efficiency, and sustainability. Water Policy 4:1–17

Roibás D, García-Valiñas MÁ, Wall A (2007) Measuring welfare losses from interruption and pricing as responses to water shortages: an application to the case of Seville. Environ Resource Econ 38:231–243

Romano G, Salvati N, Martini M, Guerrini A (2013) Water utilities and the promotion of sustainable water use: an international insight. Environ Eng Manage J 12(11):0

Ruester S, Zschille M (2010) The impact of governance structure on firm performance: an application to the German water distribution sector. Utilities Policy 18(3):154–162

Ruijs A, Zimmermann A, van den Berg M (2008) Demand and distributional effects of water pricing policies. Ecol Econ 66:506–516

Saal D, Parker D (2001) Productivity and price performance in the privatized water and sewerage companies of England and Wales. J Regul Econ 20(1):61–90

Schwarz N, Ernst A (2009) Agent-based modeling of the diffusion of environmental innovations—an empirical approach. Technol Forecast Soc Chang 76:497–511

Schwarzenbach RP, Escher BI, Fenner K, Hofstetter TB, Johnson CA, von Gunten B (2006) The challenge of micropollutants in aquatic systems. Science 313:1072–1077

Scleich J, Hillenbrand T (2009) Determinants of residential water demand in Germany. Ecol Econ 68:1756–1769

Serageldin I (2007) Water resources management: a new policy for a sustainable future. Water Int 20(1):15–21

Sibly H (2006) Efficient urban water pricing. Aust Econ Rev 39(2):227–237

Stead M, Gordon R, Angus K, McDermott L (2007a) A systematic review of social marketing effectiveness. Health Education 107(2):126–191

Stead M, Hastings G, McDermott L (2007b) The meaning, effectiveness and future of social marketing, obesity reviews, 8 (Suppl. 1), 189–193

Survis FD, Root TL (2012) Evaluating the effectiveness of water restrictions: a case study from Southeast Florida. J Environ Manage 112:377–383

Thorsten RE, Eskaf S, Hughes J (2009) Cost plus: estimating real determinants of water and sewer bills. Public Works Manage Policy 13(3):224–238

Timmins C (2003) Demand-side technology standards under inefficient pricing regimes. Environ Resource Econ 26:107–124

Tsai Y, Cohen S, Vogel RM (2011) The impacts of water conservation strategies on water use: four case studies. J Am Water Resour Assoc 47(4):687–701

United Nations (2009) Water in a changing world: the United Nations world water development. Report 3

US-EPA (2012) Innovative energy conservation measures at wastewater treatment facilities. May 2012

US-EPA (2013) Emerging technologies for wastewater treatment and in-plant wet weather management. EPA 832-R-12-011

Veleva VR (2010) Managing corporate citizenship: a new tool for companies. Corp Soc Responsib Environ Manag 17:40–51

Vörösmarty CJ, Green P, Salisbury J, Lammers RB (2000) Global water resources: vulnerability from climate change and population growth. Science 289:284–288

References

Vörösmarty CJ, McIntyre PB, Gessner MO, Dudgeon D, Prusevich A, Green P, Glidden S, Bunn SE, Sullivan CA, Reidy Liermann C, Davies PM (2010) Global threats to human water security and river biodiversity. Nature 467:555–561

Waddams C, Clayton K (2010) Consumer choice in the water sector. ESRC Centre for Competition Policy, University of East Anglia, Norwich

Wang YD, Song JS, Byrne J, Yun SJ (1999) Evaluating the persistence of residential water conservation: a 1992–1997 panel study of a water utility program in Delaware. J Am Water Resour Assoc 35(5):1269–1276

Willis RM, Stewart RA, Panuwatwanich K, Jones S, Kyriakides A (2010) Alarming visual display monitors affecting shower end use water and energy conservation in Australian residential households. Resour Conserv Recycl 54:1117–1127

Willis RM, Stewart RA, Giurco DP, Talebpour MR, Mousavinejad A (2013) End use water consumption in households: impact of socio-demographic factors and efficient devices. J Clean Prod 60:107–109

Woo CK (1992) Managing water supply shortage: interruption vs. pricing. Department of Economics and Finance, City Polytechnic of Hong Kong, working paper

Zetland D, Gasson C (2013) A global survey of urban water tariffs—are they sustainable, efficient and fair? Int J Water Resour Dev 29(3):327–342

Chapter 6
Conclusions

Scarce investments, low tariffs, and old networks and plants are some of the features of the Italian water industry that should have ended with Law 36 of 1994, when the industry began a process of intense reform that has not yet concluded. The reform brought new rules advocated by local water authorities (AATOs) and water utility firms, a review of the tariff method, and the introduction of an independent national authority (the AEEG). However, this process was often conducted without coherence or clarity, creating a huge inconvenience for both utilities and their customers. In 2012, water utilities adopted a new tariff model (the MTT), which replaced the so-called "normalized method," applied since 1996. Then, in 2013, a new method was established by the AEEG, to be applied beginning in 2014. Moreover, until June 2011, firms had charged, in their tariffs, a 7 % return on investment; however, after a national public referendum, this rate became illegal. In April 2014, water utilities were forced to reimburse citizens the unduly collected tariffs collected between August and December 2011, before the adoption of the MTT. Firms must return about 55 million euros to 11 million customers. The vague and unstable legal framework that ruled the Italian water industry is generating further damage to water utilities and, indirectly, to citizens. The uncertainty about licensing terms that emerged with Law 133/2008 (mandating the privatization of water services) and the 2011 referendum damaged the solvency of utilities and increased their financial risk; they still face many obstacles to obtaining bank loans. Consequently, efficiency and investment realization suffered under these "exogenous" conditions.

This book analyzes Italy's legal framework and then discusses three empirical surveys on the most relevant water utility management issues—efficiency, investment, and sustainability—covering the complex and troubled period between 2008 and 2012, from which emerge some interesting insights, summarized in Fig. 6.1, showing the strengths, weaknesses, opportunities, and threats (i.e., a SWOT analysis) facing the water industry.

This section was written by Andrea Guerrini and Giulia Romano.

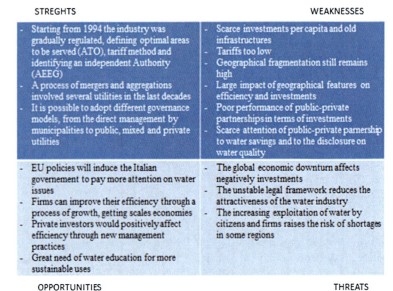

Fig. 6.1 SWOT analysis of the Italian water industry

The main strengths of the Italian water industry are threefold: first, the industry is now regulated, with a new tariff method just settled by the National Authority. The industry thus has a regulatory guide, and both the water utilities and local authorities operating throughout the country have a strong point of reference. Second, several utilities have engaged in mergers and aggregations over the last decades, so that large utilities, both mono and multibusinesses, operate in Italy; these firms adopt management best practices and are able to reach economies of scale. Finally, adopting various governance models is possible in Italy, from direct management by local government to corporatized public, mixed, or private utilities. The empirical analysis on the efficiency of water mono-utilities showed that mixed and private firms perform better than public ones do but only in terms of global efficiency: when pure technical efficiency (the capacity to purchase and consume input) is measured, ownership structure is not a significant variable. Thus, no single governance model fits all situations; it must be chosen according to factors such as the specific context, the environmental characteristics, investment needs, and the financial constraints. However, the direct provision of water services by municipalities or other local governments does not allow an easy evaluation of efficiency or performance.

After the wide-ranging reforms of the last decades, some weaknesses remain in Italy's water industry, though they have been partially reduced. Despite the economic downturn and the current uncertainty, investments and tariffs have slightly grown. The Italian unit price of household water supply and sanitation services remains one of the lowest among OECD countries (OECD 2010); the data show that the average water usage per person in Italy is the highest among European countries (OECD 1999), though household water consumption decreased after 2002 over the subsequent 10 years by around 15 % (Istat 2013). Leakages accounted for around 36 % of the water fed into Italy's water grid (OECD 2013), with a maximum of 43 % on average in the south (Cittadinanza attiva 2013). According to Eurostat data, Italy's total freshwater abstraction by public water supply is the highest among European countries.

Firm fragmentation decreased after the creation of optimal areas served (ATO) during the 1994 reform, which was followed by several mergers and acquisitions carried out in the last 20 years. However, fragmentation in the provision of water, wastewater, and sewerage services is still high; thus, Italian water utilities should continue to pursue growth to gain efficiency through economies of scale, as demonstrated in Chap. 3, even if costs are also influenced by other factors, only partially controllable by managers and policymakers, such as customer density and geographical localization. Sparsely populated and southern areas showed the greatest inefficiencies: the entrance of private shareholders, who can introduce the new management practices and techniques, usually employed by private companies, and a renewal of water mains, sewerage networks, and plants might improve this situation. However, privatization carries some negative implications. First, public–private partnership (the so-called "mixed-ownership" firms) and fully private firms realized an average € 108 of investments per capita between 2008 and 2012, less than half of the € 245 made by public water utilities. Thus, the private shareholder pays greater attention to the consumption of inputs but aims to keep investments low in order to reduce CAPEX and improve profitability in the short term. This situation also occurs as a result of *ex-ante* regulation, which allows the CAPEX to be charged on the tariff for planned investments, even if not yet realized. The new MTI tariff method should stop these practices, allowing charges on the tariff only for realized investments.

A further negative implication of private ownership is the poor promotion of sustainable water use through the reduction of household water consumption and the scarce information on water quality provided to customers. This could be explained as the effect of private owners' conflict of interest: they are maximizing profit through a continuous increase in cubic meters of water sold, and any sustainability practice conflicts with this main interest. Similarly, the lack of information on water quality is due to the scarce attention paid by private shareholders to customer needs and to the need to acknowledge the characteristics of their water. Providing information on the quality of the water provided to customers is also a means of increasing the consumption of tap water instead of bottled water, thus reducing waste.

A regulation model based on the private management of water services requires a strict set of rules to overcome information asymmetry between firms, local and national authorities, and citizens, similar to what has been done in England and Wales through the Water Services Regulation Authority (OFWAT). In this case, the Authority periodically provides on its website information on tariffs, investments, water quality, and leakages for every utility, as well as their financial statements. In Italy, the former commission of water resources (Co.N.Vi.R.I) created an information system to monitor the water utilities, but it collected little information and was not freely accessible. Even for a model where public, mixed, and private firms coexist, as with the Italian one, adopting reforms to improve market transparency is recommended to incentivize benchmarking and competition in an industry that is a natural monopoly.

The water industry now faces new opportunities. Protecting the environment and sustainably managing natural resources such as water are among the activities supported by the European Union (EU). In light of the current EU challenges concerning resource efficiency and sustainable development, the Italian government will be induced to pay more attention to water issues, thus avoiding new infringement proceedings by the EU Court of Justice for its failure to comply with water and wastewater directives.

The data show that water utility firms can improve their efficiency through growth: policymakers and water managers should therefore sustain the aggregation process in order to produce economies of scale. Moreover, the presence of private investors would positively affect efficiency through the introduction to the industry of new management practices. Finally, the empirical data on information campaigns about reducing household water consumption demonstrate that there is room for improvement in the promotion of conservation programs aiming to save water and preserve the environment, particularly in public water utilities (and even more so where local government provides the water services directly), whose shareholders (i.e., municipalities or provinces) should have the same interests as do the community, customers, and environment.

Finally, the major threats faced by the Italian water industry are linked to the global economic downturn that is negatively affecting investment opportunities and financing choices, along with the unstable legal framework that reduces the attractiveness of the water industry for private investors. The latter situation should probably improve over the next years if the government and national authorities pursue a new path and develop a stable and clear legal framework.

Even if most Europeans have historically been shielded from the social, economic, and environmental effects of severe water shortages, the gap between the demand for and availability of water resources is reaching critical levels in many parts of Europe, including Italy. Climate change is likely to exacerbate current pressures on European water resources. Moreover, much of Europe will increasingly face reduced water availability during the summer months, and the frequency and intensity of drought is projected to increase, particularly in the southern and Mediterranean countries such as Italy. Thus, the increasing exploitation of water by households, industry, and agriculture raises the risk of shortages in some regions; these must be managed carefully in order to avoid dangerous situations for citizen wellness and economic development.

References

Cittadinanza attiva (2013) Dossier acqua 2013
Istat, Istituto Nazionale di Statistica (2013) Italia in cifre. http://www.istat.it/it/files/2011/06/Italia_in_cifre_20132.pdf
OECD (1999) Household Water Pricing in OECD Countries. OECD, Paris
OECD (2010) Pricing Water Resources and Water and Sanitation Services. OECD Publishing, Paris
OECD (2013) OECD Environmental performance review Italy